Bougainvillea
A Color Handbook

About the Author

Dr. R. K. Roy, was former Sr. Principal Scientist and Head, Botanic Garden, Floriculture and Landscaping Division, CSIR-National Botanical Research Institute, Lucknow during 1989-2016. Prior to that Dr. Roy was Assistant Secretary and Horticulturist at The Agricultural –Horticultural Society of India, Alipore, Kolkata (1985-89). His main field of specialization is Ornamental Horticulture, Floriculture and Landscaping. During the tenure of his service spanning over 35 years, Dr. Roy has been deeply involved in the R&D work on Floriculture and Landscaping. He has developed 16 new varieties of ornamentals (Bougainvillea, Canna, Gladiolus and Putranjiva) which are popular in horticultural trade. Moreover, he was credited for the maintenance of authentic germplasm collection of Bougainvillea as a 'National Repository' at CSIR-NBRI. Recognizing his contribution, he was awarded with 'Dr. B.P. Pal' Award by the Bougainvillea Society of India, IARI New Delhi.

Dr. Roy is member of various professional and academic societies namely – Indian Society of Ornamental Horticulture, The Horticultural Society of India, Indian Society of Plant Genetic Resources, International Society of Environmental Botanist, Indian Association of Angiosperm Taxonomy, International Society for Horticultural Science, Belgium. He is Sr. Vice President of Bougainvillea Society of India and Vice-Chairman, Commission Landscape and Urban Horticulture, International Society for Horticultural Sciences (ISHS), Belgium.

He was also awarded with international fellowship / scholarship viz. 'RHS's Financial Award' by Royal Horticultural Society, London, United Kingdom in 1998; 'Commonwealth Science Council Fellowship', Commonwealth Science Council, United Kingdom for studying 'The Eden Project', Cornwall, United Kingdom in 2003; 'Coke Trust Award' by Royal Horticultural Society, United Kingdom for attending BGCI's 6th International Congress on Education in Botanic Gardens at University of Oxford, United Kingdom in 2006. Moreover, Dr. Roy had international exposure by way of visiting many countries and leading botanical / ornamental gardens in United Kingdom, France, Poland, Switzerland, Germany, The Netherland, Belgium etc.

He has written many books viz. 'Ornamental Annuals', 'House Plants', 'Kitchen Garden', 'Home Garden', 'Gladiolus', 'Gerbera', 'Marigold', 'Fundamentals of Garden Designing' and 'Ornamental Trees of India'. Dr. Roy is a renowned scientist in the field of Ornamental Horticulture, Floriculture and Landscaping in India and his work has also been recognized internationally.

Bougainvillea
A Color Handbook

Author

R K Roy

2019

Daya Publishing House®

A Division of

Astral International Pvt. Ltd.
New Delhi – 110 002

Published by : **Daya Publishing House®**
A Division of
Astral International Pvt. Ltd.
– ISO 9001:2015 Certified Company –
4736/23, Ansari Road, Darya Ganj
New Delhi-110 002
Ph. 011-43549197, 23278134
E-mail: info@astralint.com
Website: www.astralint.com

Printed at : **Replika Press Pvt. Ltd.**

डॉ. राकेश चन्द्र अग्रवाल
महा पंजिकार
पौधा किस्म और कृषक अधिकार संरक्षण प्राधिकरण,
कृषि एवं किसान कल्याण मंत्रालय, भारत सरकार
एन. ए. एस. सी. काम्पलैक्स, डीपीएस मार्ग,
नई दिल्ली-110012

DR. R.C. AGRAWAL
REGISTRAR-GENERAL
Protection of Plant Varieties and Farmers' Rights Authority,
Ministry of Agriculture and Farmers Welfare,
Government of India
NASC Complex, DPS Marg, New Delhi-110012
दूरभाष/Tel: +91-11-25843316 फैक्स/Fax: +91-11-25840478
Website: www.plantauthority.gov.in
E-mail: rg-ppvfra@nic.in

Foreword

Bougainvillea (*Bougainvillea* Commer.) belongs to the family Nyctageniceae and is a native of Brazil. The plant is very popular in tropical and sub-tropical gardens of the world due to colourful bracts. The popularity and use of Bougainvilleas is more in Asia particularly in India, China, Taiwan, Thailand and Malaysia. There are around 1000 varieties available in the world and every year new varieties are being added with attractive colours and foliage.

In India, many R&D institutions viz. NBRI, Lucknow; IARI, New Delhi; IIHR, Bangalore; BARC, Mumbai have been maintaining live germpalsm collection besides undertaking breeding work for the development of new varieties. Moreover, Agri-Horticultural Societies at Kolkata, Chennai and State Agricultural Universities have also been doing commendable work on Bougainvillea. Considering significant contribution made in India, IARI, New Delhi has beenrecognized as international centre of registration of new varieties. Moreover, Protection of Plant Varieties & Farmers' Right Authority, India has also formulated guidelines of registration of new varieties of Bougainvillea. Any individual, grower, farmers and nurserymen can get their new varieties registered with the Authority.

Dr.R.K.Roy, Former Sr. Principal Scientist & Head, Botanic Garden, Floriculture and Landscaping Division, CSIR-NBRI, Lucknow has written this book on Bougainvillea. The book contains comprehensive information on all aspects of Bougainvillea specially highlighting historical background, domestication, genus, species and varieties in cultivation besides characterization. In addition, a comprehensive list of varieties together with morphological characters (vegetative and floral) has been provided in order to make identification as simple as possible.

This book is an important publication considering its content of information and presentation of scientific results in simple forms which is easily graspable. I am sure students, researchers, growers, nurserymen and breeders will be benefitted.

I wish a great success of the endeavour made by Dr. Roy.

(R CAgrawal)

Preface

Bougainvillea (*Bougainvillea* Commer.) is a perennial ornamental plant grown for colouful bracts. They are very popular in ornamental gardens of tropical and sub-tropical countries as flowering plant mainly due to the colour of the bracts in various shades of white, yellow, red, orange, mauve. The strong effect produced by the bracts when in mass bloom is really adorable and a point of attraction. In tropical climate, no other plants other than Bougainvillea can influence the landscape in such a magnificent way.

At present, the popularity of bougainvillea is in peak due to its free flowering habit, low maintenance requirement, availability of newer varieties and grafted plants in multiple colour. They are highly suitable for all types of gardens and in avenue plantation also. The demand is growing every passing year in national and international market. Both professional and amateur gardeners have been using bougainvilleas so extensively, that the nursery trade is booming up.

However, the availability of literature on Bougainvillea especially on growing tips, pruning requirement, growth habit, availability of bract clours, varieties and their proper use in landscape is very low. Amateur gardeners, professionals as well as nurserymen found it difficult to know the plant in a better way together with its use depending upon the requirement. In view of that, this book was planned and written to provide all such information. This book contains 12 chapters on various aspects viz. Nativity, Historical Aspects, Migration Route and Domestication; Genus, Species and Varieties in Cultivation; Characterization; Culture and Management; Multiplication Methods, Breeding and Development of New Varieties; Classification of Varieties and their use in Landscaping. In addition, a comprehensive list of varieties together with morphological characters (vegetative and floral) has been provided in order to make identification as simple as possible. The information provided in this book is most recent and based on scientific studies coupled with practical experience of the author spanning over 35 years.

The main purpose of the book is to provide comprehensive information on all aspects of Bougainvillea. I am sure the book will serve as a reference book for all Bougainvillea growers.

I am grateful to past and present Directors of CSIR-NBRI for providing facilities for conducting studies on Bougainvillea as Scientist. My appreciation is due to my colleagues, friends and well wishers from India and abroad for their valuable advices which have made it possible to write this book.

(R.K.Roy)

Contents

1
Introduction

Bougainvillea (*Bougainvillea* Commers.) is a shrubby-climber and one of the most popular ornamental plants in tropical and sub-tropical gardens. This is grown all over the world, especially in the tropical parts, for their colourful bracts. Due to high popularity and extensive use in gardening, Bougainvillea is called 'Glory of the Tropics' in the tropical countries. The plant has travelled a lot from its place of origin - Rio De Jenerio, Brazil and domesticated in almost all parts of the world. Growth, proliferation and development of bract colour are, however, best suited to tropical and sub-tropical climate. Therefore, the popularity of Bougainvilleas is much more in the tropical parts of the world. The worldwide popularity of Bougainvillea even beat the imagination of Commerson, the French Botanist, who first collected the plant from Rio de Janerio, Brazil.

In ornamental gardening, Bougainvilleas are used extensively in various ways, particularly in Asian countries viz. China, India, Korea, Malaysia, Mauritius, Myanmar, Philippines, Singapore, Sri Lanka, Taiwan and Thailand. In Europe, due to cold climate, the use of Bougainvilleas is restricted and season bound besides needs protection from frost. Similarly, in North America, use of Bougainvilleas is very limited due to unfavourable climatic conditions required for growth and flowering. In South American countries, Bougainvilleas are commonly grown as garden plant. The other centers of popularity and ornamental use of Bougainvillea are found in tropical Australia, East and South Africa. Moreover, Bougainvillea is designated as 'National Flower' of Grenada and Zambia.

Because of their colourful bracts available in a wide range of colours *viz.* white, yellow, pink, red, mauve, bi-coloured and multi-coloured, Bougainvilleas are used extensively in comparison to other flowering garden plants. Moreover, leaves of some of the varieties have unique variegation in various colours and combinations. The colourful and attractive foliage is also an added attraction particularly when the plant is out of bloom.

The high adaptability of Bougainvilleas to different agro-climatic conditions and free flowering habit, have made them an obvious choice for every garden. A large number of varieties are available in the horticultural nursery trade with varied growth habit and ideal for growing in the gardens for various purposes. The plant withstands well in adverse climatic and stress conditions due to hardy nature and produce colourful bracts as well. The requirement of maintenance is also less in comparison other garden plants. Considering overall performances, Bougainvilleas are most suitable flowering plant for the gardens of sub-tropical and tropical regions of the world. No other plant can influence the landscapes and gardens in such an incredible way.

Bougainvilleas are extensively used in ornamental gardening in India, beside as a plant for decoration of avenues also. However, the season of flowering and intensity vary with a great magnitude. Bougainvilleas adorn every region of India with their colourful bracts and mass blooming. Southern India comprising of Bangalore, Mysore, Chennai, and Hyderabad have favourable agro-climatic seasons. Profuse blooming takes place during February to April and August to October in succession and profusion. Similarly, Pune, Nasik, Nagpur, Bombay regions of Western India usually have extended blooming in flashes round the year due to the prevailing moderate climate in that region. As a whole, eastern, western and southern India usually have extended blooming period in comparison to northern region. Mass flowering of Bougainvilleas in northern India (Delhi, Chandigarh, Patiala, Agra, Lucknow, Kanpur) and in adjoining areas usually takes place during March to May followed by pre-winter blooming in November-December. During very cool period (December end to January) due to low temperature and less availability of sunshine, Bougainvilleas remain almost dormant. The performance of Bougainvilleas with regard to growth and flowering is also appreciable in hilly regions also. They can be grown at a height of 1500-2000 meter above the sea level. Good flowering has been found in Solan, Shimla, Almora, Nanital and other northern hilly areas.

Irrespective of the season, the lovely colourful bracts of Bougainvilleas in profusion always draw attention of the visitors. In harsh climate when other ornamental plants fail to perform, Bougainvilleas produce flowers abundantly. The oddest garden corner where no other plant survives, Bougainvilleas gracefully adorn. The range of bract colour, massiveness of flowering and intensity are truly amazing which change the face of garden into a vivid landscape.

By virtue of their striking colourful effect of the bracts together with variegated foliage, every garden lover makes Bougainvillea an obvious choice for their garden.

*****

2

Nativity, Historical Aspects, Migration Route and Domestication

Nativity and Historical Aspects

The nativity of Bougainvillea dates back to 1768 A.D. The plant was discovered by Dr. Philibert Commerson (1727-1773), a French explorer and naturalist in Rio de Janerio, Brazil. In 1766, a ship called La Boudeuse sailed from Nantes on a round the world voyage (1766-69) which was commissioned by the French Government. Louis Antonie de Bougainville (1729-1811), a mathematician and an admiral was in command of the ship. The ship reached Rio de Janerio and the landing made a history in Horticulture so far as flowering plants are concerned and Bougainvillea in particular.

The newly collected plant was named after Louise Antonie de Bougainville, a close friend and admiral of the ship by Dr. Commerson. After 20 years of Commerson's discovery, the genus name *Bugainvillea* was appeared in Genera Plantarum by A.L. de Jusseru in 1789. The generic name underwent several transformations and finally corrected and adopted as *Bougainvillea* by Spachs (1841) which was subsequently published in the Index Kewensis (Supp. 9: 1931-35).

Two species, namely *Bougainvillea spectabilis* and B. glabra were initially introduced in the early 19th century from place of origin to Europe. Contemporary to this, there was another important event took place in the history of Bougainvillea. The discovery of crimson Bougainvillea in Cartagaea, a Spanish port in the Mediterranean was made by Mrs. R.V. Butt. It was thought to be a distinct species but later found to be natural hybrid between *B. glabra* and *B. peruviana*. That was named after Mrs. R.V. Butt as 'Mrs. Butt'. Thereafter, occurrence of natural hybrids all over world was common. The main basal true species when grown together yielded many hybrids spontaneously in East Africa, Canary Island, Australia, North America, Philippines and India.

Migration Route

After its discovery, Bougainvillea travelled a lot and migrated to Europe as a primary centre of introduction and domestication during early 19th century. However, the route, period and centre of introductions were not exactly known.

Europe - *B. spectabilis* was first introduced to France in 1829 from Peru as per report published in Paxton's Botanical Magazine. The plant flowered successfully in Paris around 1835. Subsequently, *B. spectabilis* was introduced to United Kingdom (Great Britain) from southern Brazil in 1844. However, the newly introduced plant could not flower and the attempt of its introduction and domestication remain unsuccessful. Significant individual effort was made by J.D. Damiels, Thames, England. He got profuse flowering in a container grown plant and that arose a lot of interest among the plant lovers. The popularity of Bougainvilleas as a garden plant grew slowly. Around 1860, other species of Bougainvillea namely *B. glabra, B. peruviana* were introduced to United Kingdom.

USA - The domestication of Bougainvillea in Florida, USA dates back 1881. Remarkable individual effort was made by a nurseryman - Pliny Reasoner (Reasoner's Tropical Nurseries) long before establishment of Plant Introduction Bureau, USDA. 'Splendens' was the first Bougainvillea variety introduced and grown in Florida which was brought from Havana in 1885. With that introduction and popularization, several other varieties of *B. spectabilis*, *B. glabra* were introduced, multiplied and sold to the garden lovers. *B. glabra* 'Sanderiana' was introduced to the plant lovers in 1894 when that was exhibited in the flower show at Royal Horticultural Society, London by F. Sander & Co. Though the variety was from old English Garden but originally introduced from South America. Subsequently, this variety was exported to Singapore in 1894. Bougainvillea 'Splendens' was also exhibited in London in 1861.

Development of new varieties paved the way for increasing the popularity of Bougainvillea. Many other varieties were introduced and developed as a result of cross breeding done by the amateur growers and nurserymen. Some notable varieties were - 'Afterglow' (orange), 'Crimson Lake' (crimson), 'Helen Coppinger' (purplish rose-pink), 'Panama Pink' (soft pink), 'Rosa-Catalina' (rose), 'Refulgens' (purple), 'Lateritia' (mauve) etc.

Asia - In Asian countries, migration and introduction of Bougainvillea started around 1800 A.D. The main centers of introduction and domestication were - Philippines, Mauritius, India, Singapore, Taiwan, China and Malaysia. Some of the varieties were migrated directly from South American countries while others via England. Initially, amateur plant lovers, travelers, colonial civil servants and their family members introduced several varieties. As per reports, *B. glabra* 'Sanderiana' was exported to Singapore by Sanders & Co., Florida, USA in 1894. Similarly, *B. glabra* was introduced to Mauritius in 1860 and subsequently brought to Calcutta in 1869.

The most significant role in the introduction and domestication of Bougainvilleas and other ornamental plants in India was played by the Agricultural and Horticultural Societies established by British Government. Very specifically, the Agricultural and Horticultural Society of India (AHSI) at Alipore, Calcutta played a pioneering role. *B. spectabilis* was introduced to India from Royal Botanic Garden, Kew by this society. Thereafter, improvement work on Bougainvillea was also started in this society by renowned British horticulturist Mr. S. Percy Lancaster. He was credited with the development of first Bougainvillea variety, 'Scarlet Queen Variegata,' in India in 1926. Introduction of another variety 'Mrs. Butt' from Royal Botanic Garden, Kew to the Agri-Horticultural Society of India (AHSI) in 1923 created sensation and paved the way for further popularization of Bougainvillea in India. Commendable effort was made by Mr. S. P. Lancaster together with development of the new variety - 'Mary Palmer' helped popularization of Bougainvillea in different parts of the country. A new variety named as 'Princess Margret Rose' was developed by the Agri-Horticultural Society, Madras in 1935, and further popularized Bougainvillea in India.

Lalbagh Botanic Garden, Bangalore also did a commendable job on introduction and development of new varieties very similar to the work done at the AHSI, Alipore, Calcutta. Some remarkable new varieties was introduced there from Kenya, Africa were - 'Isabel Greensmith', 'Asia', 'No. 2', 'Elizabeth', 'Kayata', 'Closeburn'. In addition to above, the other were - 'Glady's Hepburn', 'Natalli' (Durban, South Africa), 'Mahara', 'Roseville's Delight' etc. Moreover, new varieties were developed at Lalbagh Botanic Garden by exploiting the existing germplasm collection. Some of them were - 'Trinidad', 'Raman', 'Gagarin' etc.

Popularization and Development of New Varieties in India and Abroad

Introduction and popularization of Bougainvillea was initiated in India with the import of Bougainvillea spectabilis by AHSI, Calcutta, in 1869 from RBG, Kew. Lalbagh Botanic Garden, Bangalore was another important centre of introduction. A group of varieties were imported from Kenya, Durban and domesticated in 1961. The remarkable ones were 'Isabel Greensmith', 'Asia', 'Elizabeth', 'Closeburn', 'Natalii', 'Gladys Hepburn' etc. Similarly, other Agri-Horticultural Societies and Botanic Gardens in India were also initiated effort for the introduction of new varieties, multiplication, display and popularization.

Significant role was played by several horticultural societies, individuals, nurserymen, institutions for the development of new varieties. The credit for the development of first Bougainvillea variety ('Scarlet Queen Variegata') goes to Mr. S. Percy Lancaster at AHSI, Calcutta in 1926. The other prominent breeding work and varieties developed were - 'Princess Margaret Rose' (AHSI, Madras, 1935), Mary Palmer (AHSI, Calcutta, 1949); 'Dr. B.P. Pal' (NBRI, Lucknow, 1969); 'Fantasy' (B. Rama Rao, Madras); 'Louise Wathen' (AHSI, Madras, 1932); 'Alick Lancaster' (AHSI, Calcutta, 1930); 'Scarlet Glory' (K. Gopalaswamienger & Sons, Bangalore, 1952); 'Mrs. H.C. Buck' (Soundarya Nursery, Madras, 1930) etc.

The work for development of new varieties was also started with the help of hybridization, selection and bud sports. Several varieties had arisen from seedling selection as a result of natural crossing. Some artificial hybridization and subsequent development of new varieties was also done. First controlled cross pollination was done by Jim Hendry, Florida in 1927 between 'Rosa Catalina' (male) and 'Lateritia' (female).

Later, two excellent hybrids were developed and named as 'Margaret Bacon' (lavender rose-pink) and 'Daniel Bacon' (dark purple-pink). Another report from Peru mentioned two new hybrids made by W. N. Sands. He raised 'Lady Seton James' (rose) as a cross between 'Sanderiana' and 'Lateritia' followed by 'Lady Watls' (terracotta to selmon pink) as a result of cross between 'Rosa Catalina' x 'Lateritia'. Similarly, another new hybrid 'Barbara Karst' was developed which was predominantly available and used in Florida, California and South Texas.

At present Bougainvilleas are quite popular in Australia. However, the History, domestication and development was not very clearly documented. Outstanding work on collection, popularization and development of new varieties was done by Jan and Peters Iredell, Brisbane, Queensland. As a result a series of varieties, known as 'Bambino' has been developed as early as 1997. These varieties are naturally dwarf, less thorny and floriferous. Therefore, they are highly suitable for pot culture, hanging basket and in several other ways. Some outstanding varieties of this series are - 'Bluey', 'Jezebel', 'Jazzi', 'Jellibene', 'Majik', 'Panda', 'Zuli', 'Zuki' etc.

Philibert Commerson

Louis Antoine, Comte de Bougainville

'Bambino Majik'

'Elizabeth'

'Glabra Sanderiana'

'Isabel Greensmith'

'Lateritia'

'Mahara'

'Mary Palmer'

'Mrs. Butt'

'Roseville's Delight'

'Scarlet Queen Variegata'

'Barbara Karst'

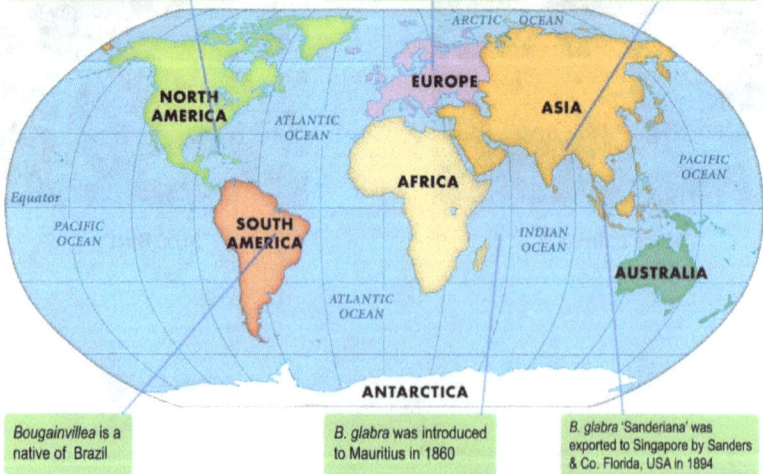

Bougainvillea introduced and grown in Florida was 'Splendens' brought from Havana in 1885

B. spectabilis was first introduced to France in 1829 from Peru

B. spectabilis was introduced in AHSI, Calcutta, in 1869 from RBG, Kew

Bougainvillea is a native of Brazil

B. glabra was introduced to Mauritius in 1860

B. glabra 'Sanderiana' was exported to Singapore by Sanders & Co. Florida, USA in 1894

Figure 2.2. Migration route of *Bougainvillea* to different countries.

*****

3

Genus, Species and Varieties in Cultivation

The placement of the genus *Bougainvillea* is given below, as per its correct taxonomical standing and systematic position.

Kingdom	:	Plantae
Infradivision	:	Angiosperms
Phylum	:	Magnoliophyta
Class	:	Magnoliopsida
Order	:	Caryophyllales
Genus	:	*Bougainvillea*

Genus

Bougainvillea Commers.

The genus name, *Bougainvillea* Commers. (Nyctaginaceae) was first coined by Philibert Commerson and published by A. L. De Jussieu in 'Genera Plantarum' in 1789. The generic name, initially, was spelt as '*Beuginvillea*', but finally accepted as '*Bougainvillea*' in honour of the famous French navigator - L. A. De Bougainville. This was as per approval of the International Code of Botanical Nomenclature and was published in Index Kewnensis (Suppl. 9: 1931-35).

Description

Shrubs or small trees, often scandent, usually armed with simple or branched spines, straight or curved; leaves alternate, petiolate, entire; flowers perfect, either solitary and subtended by 3 bracts, or usually in a 3-flowered axillary inflorescence consisting of 3 large, persistent, often brightly coloured bracts, a flower borne on the inner surface of each bract, pedicel confluent with costa of the bract; perianth tubular, limb usually shallowly 5-lobed, tube subterete or 5-angled; stamens 5-10, somewhat unequal, connate at the base into a short cup; anthocarp fusiform, coriaceous, 5-costate.

Species

Heimerl (1900) recognized all together 10 species. Three of which (*B. spectabilis* Willd., *B. glabra* Choisy and *B. peruviana* Humb. and Bonp.) were of horticultural importance having showy and colourful bracts. He noted a great deal of variation with regard to colour and shape of the bracts in *B. glabra* and *B. spectabilis*.

Standley (1931) described the following eight species under the genus *Bougainvillea* with following distinguishing key characters.

Key Characters of the Species

Bracts brightly coloured, purplish red or bright red, retaining the colour when dried, mostly 2.5-4.0 cm long.

Perianth tube glabrous ....   1. *B. peruviana*

Perianth tube variously pubescent.

Perianth tube hirsute or villous; leaves usually copiously villous..... 2. *B. spectabilis*

Perianth tube puberulent or glabrate; leaves glabrate.......   3. *B. glabra*

Bracts green, or sometimes brightly coloured when fresh but losing their colour when dried; usually smaller, 1.0-2.5 cm long.

Perianth 6.5-7.0 mm long, gradually dilated from base to apex, glabrate, or hirtous above; bracts 7-12 mm long...........................   4.*B. campanulata*

Perianth 9-20 mm long, constricted above; bracts 13-27 mm long.

Perianth glabrous, 11-14 mm long; bracts 25-27 mm long, 13-15 mm wide, glabrous except on the costa....................................   5. *B. berberidifolia*
Perianth variously pubescent; bracts 13-22 mm long.

Perianth 9-11 mm long, densely tomentulose............... .   6. *B. praecox*

Perianth 12-20 mm long.

Perianth puberulent ....   7. *B. stipitata*

Perianth pubescent or hirsute ....   8. *B. infesta*

On the basis of description and study of the morphological characters of the specimens, *B. spectabilis* Willd., *B. glabra* Choisy and *B. peruviana* Humb. & Bonp. are the only species possessing large, colourful and showy bracts which are horticulturally important for gardening throughout the world. It has also been reported that a number of different forms, having other shape, size and colour of bracts, has arisen in course of cultivation as a result of spontaneous mutation and hybridization among the basal species.

The major morphological characters of the species comprising of vegetative and floral parameters are furnished below.

1. *Bougainvillea buttiana* Holttum & Standl. **Nativity** - Brazil

Morphological description of this plant was based on specimen of *B. x buttiana* 'Mrs. Butt' brought by Mrs. Butt from Cartagena (Colombia) to Trinidad in 1910. The name 'Mrs. Butt' was first published in the Report of Department of Agriculture, St. Vincent, in 1916-17, by R. O. Williams (Holttum 1955b). As the specimen differed from all other Bougainvilleas in cultivation with regard to some morphological features including colour of bracts, Holttum and Standley (1947) described it as a separate species - *B. buttiana*. However, subsequent studies and considering hybrid nature of this plant a change in the name was necessary and Holttum (1955b) finally referred the species as *B. x buttiana*. The morphological characters of this species resemble with *B. peruviana*.

2. *Bougainvillea glabra* Choisy **Nativity** - Brazil, Venezuela, Colombia, Ecuador

A woody climber, large, branchlets puberulent at juvenile stage but glabrate subsequently; armed with numerous stout spines; leaf blades broadly ovate to ovate-lanceolate, 4.0-10.0 cm long, base rounded to acute, apex abruptly or gradually acute to acuminate, puberulent when young but turns glabrate quickly; bracts purplish-red, broadly ovate to oval, base subcordate generally, abruptly acute or acuminate at the apex, sparsely puberulent or glabrous; perianth 15-25 mm long; stamens 8; anthocarp turbinate, 7-13 mm long with 5 acute angles.

This species is commonly grown as garden plant for colourful bracts. As per Hooker's opinion, this species is closely related to *B. spectabilis* and he treated it as a variety of *B. spectabilis*. However, Standely recognized *B. glabra* as separate species.

3. *Bougainvillea peruviana* H. & B. PI. (Syn. *Tricycla peruviana* Poir.) **Nativity:** Peru, Colombia

The description of the species was prepared first by Humboldt and Bonpland (1808). The plant is shrubby in habit, 3-7 m in height, erect or scandent; branches sparsely puberulent or glabrate, spines 1.0-2.5cm long, slender, numerous; leaves thin, slender-petiolate, blades broadly ovate to suborbicular, 5-7 cm long, base subtruncate, apex subobtuse or abruptly acute, glabrous or nearly so when very young; bracts bright rose, 1.5-3.5 cm long, apex obtuse or rounded, glabrous except along the puberulent costa; perianth 16-20 mm long, white or whitish, the limb 5-6 mm broad; stamens usually 6; anthocarp about 10 mm long and 2.5 mm thick, glabrous.

4. *Bougainvillea spectabilis* Willd. **Nativity:** Brazil, Argentina, Bolivia.

Willdnow for the first time described this species in fourth edition of Linnaeus's 'Species Plantarum' (1798).

A woody vine, growth habit large, branches copiously fulvous-villous, rarely glabrate, spines numerous, stout, about 4 cm long; leaf blades broadly ovate to sub orbicular or rounded-oval, 5-10 cm long, base round to acute, often short-decurrent, apex abruptly acute or acuminate, usually densely villous beneath; bracts purplish-red, ovate-oval or broadly ovate, 2.0-4.5 cm long, base sub-cordate, apex abruptly acute or acuminate or sometimes obtuse, sparsely puberulent or short-villous; perianth 15-30 mm long, tube green, limb 6-7 mm wide, yellowish; anthocarp 11-14 mm long, densely short-villous.

5. *Bougainvillea berberidifolia* Heimerl **Nativity** - Bolivia

Shrub in growth habit, at young stage branchlets sparsely tomentulose, soon glabrate; spiny, about 1.0-2 .0 cm long; leaf blades elliptic to obovate-elliptic, broadest at or above the middle, 1.5-2.5 cm long, apex rounded or obtuse, base attenuate, glabrate; bracts bright rose, turning greenish with the maturity, ovate-oblong, 25-27 x 13-15 mm, sub-obtuse, sometimes sub-orbicular and broadly rounded at the apex, glabrous except along the costa; perianth 11.0-14.0 mm long, yellowish or red, glabrous; limb 4-5 mm broad; stamens usually 5.

6. *Bougainvillea campanulata* Heimerl **Nativity** - Bolivia, Argentina

A small tree or shrub, branchlets glabrous or tomentulose minutely when young, usually unarmed; leaves short-petiolate, elliptic-lanceolate or narrowly oblong-elliptic, 20 x12 mm in size or smaller, broadest at the middle, apex obtuse, base acute, minutely hirtellous beneath along the costa, otherwise glabrous; bracts elliptic to obovate-elliptic, 7.0-12.0 x 4.0-6.5 mm, sub-obtuse to rounded at the apex, yellowish green, sparsely puberulent or villosulous; perianth 6.5-7.0 mm long, yellow, hirtous above, limb 9-10 mm broad; stamens 7-8.

7. *Bougainvillea infesta* Griseb. **Nativity** - Bolivia, Argentina

Shrub, about 2 m in height, young branchlets densely pubescent, older glabrate; unarmed or furnished with stout spreading spines, 5.0-15.0 mm long; leaves petiolate, elliptic to ovate or elliptic-lanceolate, mostly 2-4 cm long, broadest at or below the middle, obtuse to acute, truncate to acute at the base, densely puberulent or tomentulose beneath; bracts greenish, about 20.0 x 11-13 mm, base rounded or subcordate, apex very obtuse or rounded, densely pubescent; perianth 12-14 mm long, densely pubescent or hirsute, limb 5 mm wide; stamens 5.

8. *Bougainvillea malmeana* Heimerl

A shrub in growth habit, branches partly elongated, obliquely, bearing reduced flowers; bracts smaller, coloured but not spectacularly. Leaves are lacking. Inflorescences partial typically tri, biflora, peduncle weak, 5-8 mm long , bracts pale yellow, elliptic to ovate – elliptic, 13-19x 8.5-13 mm in size, base rounded to a little subcordatae, apex obtuse to antice obtusiusculae. Perianths 9-10mm long, green - yellow, tube of the flower 2.5 mm broad, light prismatic, slender, tube concolour, the limb 6-7 mm broad; stamens 6 (rarely 5), filaments(6-9 mm, anthers 1 mm long and broad.

9. *Bougainvillea modesta* Heimerl

The species is having growth habit like a tree, reaching up to 25 m height, but unarmed, strong, and partly with elongate branches obliquely spreading branches, partly leafless flowering time, abundantly flowering, flowers bracts smaller and less notable in colour. Branches rarely little flexible, internodes greatly varying length (1-3 cm). Leaves elliptic to ovate-elliptic, 50-80 mm: 30-40 mm, tomentose, minutely acuminate, the base narrowed, somewhat obtuse to bluntly rounded, abaxial rather acute, green to grayish, with short hair on both sides. Inflorescences typically tri- partial on either side biflora   (this is also the case of bi -bracteata), peduncle weak, at flowering time of 10-14 mm long, subdensely tomentello partly erect or partly obliquely. Bracts (dry) pale yellow, little shorter, but distinct, elliptic,15-17 x 8-9 mm), base obtuse, apex acute, very short hair, a moderately densely puberulous on both sides, slender and densely nerved pale, thin, membranous; pedicel 5-6 mm. Perianths 10-11 mm long, green-brown. Stamens 6, filaments 8-9 mm long , anthers 1 mm long and broad. Ovary oblong - ellipsoid (2.5-3.5 mm long, 1.0-1.5 mm wide).

10. *Bougainvillea pachyphylla* Heimerl ex Standl.

Generally, a shrub or small tree, scandent sometimes, young branchlets finely puberulent, armed with numerous spines; leaf blades ovate to broadly elliptic, thick and somewhat leathery, 3.0-6.5 cm long, acute or acuminate or gradually narrowed to an obtuse apex, base broadly rounded, finely and densely puberulent; bracts rose-coloured, broadly elliptic-ovate to almost orbicular, glabrate; perianth densely puberulent, 9-11 mm long.

11. *Bougainvillea pomacea* Choisy [Syn. *Bougainvillea glabra* var. *Pomacea* (Choisy) Luetzelb.]

The species is closely related with *B. spectabilis*. A shrub or a tree, branches teretes, woody, glabrate, thorny; spines short, green, strong curved, glabrate; leaf blade alternate, petiolate, ovate-elliptic or obtuse, slightly undulate-crenulate, petiole 3 cm long, glabrate; inflorescence axillary, peduncle divided into 2-3 short branches, puberulent; bracts papery, elliptic-ovate, base sub-cordate, 4-6 cm long, large, blood red colour, membranous veins reticulate.

12. *Bougainvillea praecox* Griseb. (Syn. *Bougainvillea praecox* var. Rhombifolia Heimerl; *B. praecox* var. spinosa Chod. & HassI. Bull. *; B. praecox* var. rhombifolia Heimerl, *B. modesta* Heimerl, Denkschr. Akad. Wiss. Wien) **Nativity** - Bolivia, Argentina and Paraguay

Shrub or tree in growth habit, branchlets tomentulose or turns glabrate; sparsely spiny or unarmed, spines generally 5 mm long or less; leaf blades ovate to elliptic or ovate-elliptic, 8.0 x 4.0 cm or smaller, narrowed to an obtuse apex, acute or attenuate at the base, glabrate; bracts white or reddish, turning greenish when withered, broadly ovate, about 14 x 12 mm, base rounded or subcordate, apex rounded or very obtuse, densely puberulent, especially on the veins; perianth 9.0-11.0 mm long, densely tomentulose, limb 4 mm broad; stamens 5-6.

[The two varieties ('Spinosa' and 'Rhombifolia') described by Heimerl are minor forms and not worthy of nomenclatorial recognition. He also suggested the possibility of combining *B. praecox* and *B. modesta* due to insufficient grounds to separating them. *B. modesta* was described as a tree but the description was probably erroneous and sufficient diagnostic characters were not found].

13. *Bougainvillea spinosa* (Cav.) Heimerl (Syn. *Bougainvillea patagonica* Decne.)

Usually a shrub, reaching 2-4 m height, branches stout sparsely puberulent on the younger parts; armed with stout, rigid spines, 2 cm long or less, furcated at the apex; leaves linear-spatulate or oblong-spatulate, 9-15 mm long, 2-4 mm wide, thick and fleshy, obtuse or rounded at the apex, gradually narrowed to the base, very shortly petiolate, glabrous; flower borne on a slender peduncle 3-6 mm long, solitary; bracts green, shorter than the flower, broadly cordate-ovate to orbicular, about 12 mm long, sparsely puberulent; perianth 8-13 mm long, glabrous or sparsely puberulent; anthocarp 6-7 mm long.

14. *Bougainvillea stipitata* Griseb. (Syn. *Bougainvillea frondosa* Griseb.)
Nativity - Bolivia, Argentina, Brazil

(Syn. *B. longispinosa* Rusby, Mem. ; *B. stipitata* var. longispinosa Heimerl, Denkschr. Akad. Wiss. Wien ; *B. stipitata* var. Kuntzeana Heimerl ; *B. stipitata* var. Fiebrigii Heimerl.)

Shrub or small tree; young branches puberulent or tomentulose, unarmed or armed with stout or slender spines 2.5 cm long or less; leaf blades ovate, rhombic, deltoid-ovate, or rarely lanceolate, usually acuminate but sometimes obtuse, base rounded to acute, puberulent or glabrous, 3.5-7.5 cm long; bracts green or tinged with rose, greenish when dried, broadly ovate or ovate-elliptic, 14-20 x 12-15 mm wide, base subtruncate to shallowly cordate, apex obtuse or acutish, sparsely or rather densely puberulent; perianth 13-20 mm long, greenish, puberulent or rarely glabrous; stamens 7-8; anthocarp fusiform, 12-15 mm long, minutely puberulent or glabrous.

Other recognized species of Bougainvillea are given below but no morphological description of the species is available.

15. *Bougainvillea herzogiana* Heimerl

16. *Bougainvillea lehmanniana* Heimerl

17. *Bougainvillea lehmannii* Heimerl

18. *Bougainvillea trollii* Heimerl

Varieties

Large numbers of varieties are available all over the world. However, they remained restricted within the geographical zones / countries and not much available for horticultural purposes throughout the world. Moreover, there is no comprehensive of list available. R&D institutions and nurserymen though maintained catalogue for the varieties, often that was not a complete list. Therefore, a comprehensive list of Bougainvillea varieties available all over the world has been prepared on the basis of available information (Table 3.1). This will facilitate popularity and exchange of germpalsm collection from one country to other.

Table 3.1. A comprehensive list of varieties available all over the world:

SL. No.	Name of the Variety	Breeder / Raised or Introduced by and Year
1	'Abimanyu'	B. K. Banerji, NBRI, Lucknow, India, 2010; Bud Sport.
2	'Abraham Kavoor'	-
3	'Ada Atwil'	Rodney Jonklaas, Jaela, Sri Lanka.
4	'African Sunset'	-
5	'Afterglow'	Florida, U.S.A (Reported by Eric V. Golby); Bud sport.
6	'Aida'	S. Percy Lancaster, Calcutta, India (1935-40); Introduction.
7	'Aida Variegata'	NBRI, Lucknow, India; bud sport.
8	'Alba'	Mr. H.P. GreenSmith, Bangalore, India, 1961; Introduction
9	'Alex Butchart'	W.F. Turley, Australia; Hybrid Seedling.
10	'Alicia' (white)	-
11	'Alick Lancaster' (Syn. 'Lilac Queen')	A. Percy Lancaster, Delhi (1930), India.
12	'Alizon Davy'	-
13	'Amaranth'	S. Percy Lancaster, 1931.
14	'Amarault'	S. Percy Lancaster, 1931;Introduction.
15	'Ambience' ('James Walker')	-
16	'Anabello'	Nairobi, Kenya, East Africa
17	'Anne Faed'	Mr. Ruth Faed, Trelawney, Rhodesia.
18	'Apple Blossom'	-
19	'Apricot Dream'	Introduction
20	'Archana'	G.S. Srivastava, N.B.R.I., Lukcnow, India, 1975.
21	'Ardenxi'	Introduction
22	'Arjuna'	M.N. Gupta & R. Shukla, N.B.R.I., Lukcnow, India,1974.
23	'Aruna'	S.C. Sharma & R.K.Roy, NBRI, Lucknow, India, 2008.
24	'Ashley Nagpal'	Vema Nagpal Bombay, India, 1986.
25	'Asia'	Mr. H.P. GreenSmith, Bangalore, India, 1961.
26	'Assorted'	Introduction
27	'Audrey Davision'	Mrs. D.A. Delap, Kenya, Africa.
28	'Audrey Delap'	Mrs. D.A. Delap, Kenya, Africa.
29	Aurantiaca (Syn. Lindleyana)	Canary Island.
30	'Aussie Gold' (Yellow)	-
31	'Autumn'	S. Percy Lancaster, Calcutta, India (1935-40).
32	'B.G.K. Hamilton'	A.W. Burnely, Kenya, Africa
33	'B.J.F. Bailey'	Kenya, Africa
34	'Baby Margaret Rose'	A.P.A.U.
35	'Bagen Beauty'	Introduction
36	'Bambino Baby Alyssa'	Australian Variety
37	'Bambino Baby Lauren'	Australian Variety
38	'Bambino Baby Victoria'	Australian Variety
39	'Bambino Beesnees'	Australian Variety

40	'Bambino Bokay'	Australian Variety
41	'Bambino Jazzi'	Australian Variety
42	'Bambino Majik'	Australia Variety
43	'Bambino Maudi'	Australian Variety
44	'Bambino Miski'	Australian Variety
45	'Bambino Nonya'	Australian Variety
46	'Bambino Panda'	Australian Variety
47	'Bambino Pedro'	Australian Variety
48	'Bambino Shaba'	Australian Variety
49	'Bambino Siggi'	Australian Variety
50	'Bambino Zuki'	Australian Variety
51	'Bambino Zulu'	Australian Variety
52	'Beryl Lemmen'	-
53	'B.T. Red'	Thailand
54	'Barbara Karst'	J.E. Hendry, Everglades Nursery, Florida U.S.A, 1927
55	'Barnabee'	J.E. Hendry, Everglades Nursery, Florida, U.S.A, 1927
56	'Baron Hado'	B.A Rama Rao Madras, India
57	'Beauty'	S. Percy Lancaster, Calcutta, India (1935-40)
58	'Beba' (Pink)	-
59	'Begum Ali Yawar Jung'	Habibur Rehman, Aligarh, India
60	'Begum Sikander'	S.N. Zodoo & T.N. Khoshoo, N.B.R.I., Lucknow, India, 1969
61	'Beryl's Red'	
62	'Betty Hendry'	J.E. Hendry, Everglades Nursery, Florida, U.S.A., 1927
63	'B.G.K. Hamilton'	A.W. Burnely, Kenya, Africa
64	'Bhabha'	Lalbagh Botanic Gardens, Bangalore, India,1960
65	'B.J.F. Bailey'	W.F., Turley, Queensland, Australia
66	Blondie (Syn. Hugh Evans)	Originated in Kenya, Africa
67	'Blueberry Ice'	New Introduction
68	'Boise-de-Rose'	W.F. Turley, Queensland, Australia
69	'Bombay Burbundy'	Introduction
70	'Bonfire'	S. Percy Lancaster Calcutta, India (1935-40)
71	'Brasiliensis'	R.E. Holttum, 1970
72	'Bridal Bouquet' (White)	-
73	'Brilliance'	
74	'Brillant'	Mr. H.P. Greensmith, Kenya, Africa
75	'Brilliant Variegata'	Lalbagh Botanic Gardens, Bangalore, 1975
76	'Bronda Ohare'	Verna Nagpal Bombay, India, 1984
77	'Buddhadas'	M. Buddhadas, Yediur Botanic Gardens Bangalore, India
78	'California Gold'	Introduction
79	'Camarillo Fiesta'	Introduction
80	'Cannellii'	M/s Cannell & Sons, England, 1904.
81	'Cardinal'	Kenya, Africa

82	'Carloton Corea'	-
83	'Carmencita' (Syn. Glasalan Red)	Philippines
84	'Carnarvon'	Carnarvon, Western, Australia, 1935.
85	'Carpet Pink'	Introduction
86	'Carpet Purple'	Introduction
87	'Carpet White'	Introduction
88	'Carnival'	Introduction
89	Cascade	Agri-Horticultural Society, Madras. India
90	'Celia Braganza'	Verna Nagpal, Bombay, India, 1986
91	'Cerise Sunset'	Introduction
92	'Ceylon Hybrid'	Introduction, Ceylon, 1913
93	'Chandrabieri'	M/s Chandra Nursery, Sikkim, India
94	'Charles William'	Origin Jamaica, Imported into Kenya by H.A. Delap, 1951.
95	'Charles Wilson'	Introduction from Kenya at Lal Bagh Botanic Garden, India, 1961.
96	Cherry Blossom (Syn. 'Bridal Bouquet', 'Double White', 'Mahara off white')	Reported by Dr. J.V. Pancho (1967), College of Agriculture Laguna, Philippines
97	'Cherry Ripe'	S. Percy-Lancaster, Calcutta (1935-40).
98	'Chinese Craker'	-
99	'Chitra'	T.N. Khoshoo, D.Ohri and S.C. Sharma, NBRI, India, 1981.
100	'Chitravati'	IIHR, Bangalore, 1979.
101	'Cindrella'	Jan Ire dell, Australia, 1986.
102	Claret Rose	Jan Iredell, Australia, 1986
103	'Cleopatra'	-
104	'Clifton Carnival'	New Introduction
105	'Closeburn' (Syn. 'Temple Fire')	Introduction from Kenya, Lal Bagh Botanic Grden,1961.
106	'Closeburn Cardinal'	H. Grahame Bell, Kenya, Africa, 1930.
107	'Coconut Ice'	Ian Iredell, Australia, 1986.
108	'Coconut Pink'	-
109	'Colour Splash'	Rodney Jonklaas, Jaela, Ceylon
110	'Coker'(Syn. Penang)	R.E. Holttum, Singapore, 1938
111	'Common Rose'	S. Percy-Lancaster, Lucknow, India,1959
112	'Conquest'	S. Percy-Lancaster, Calcutta, India (1935-40)
113	'Constance'	S. Percy-Lancaster, Calcutta, India (1935-40)
114	'Convent'	M/s Evans & Reeves, Los Angeles, U.S.A
115	'Copper King'	New Introduction
116	'Crimson Jewel'	U.S.A.
117	'Crimson King'	M/S Soundarya Nursery, Madras, India
118	'Crimson Lake'	S. Percy-Lancaster, Calcutta, India (1935-40)'
119	'Crimson Lake Jr.'	J.E. Hendry, Ever glades Nursery, Florida. U.S.A., 1927
120	'Crimson Red'	-

121	'Crimson Yellow'	Rodney Jonklaas, Jaela, Ceylon.
122	'Crispa'	Thailand
123	'Cypheri'	Exhibited and named at Shewsbury, England, 1897
124	'Daeng Ban Yen'	New Introduction
125	'Daniel Bacon'	J.E. Hendry, Everglades Nursery, Florida, U.S.A., 1927
126	'Daphene Mason'	Greensmith, Nairobi
127	'Dar-es-Sallam' (Syn. Lateritia)	Exhibited in London, 1865; introduced to india in 1883
128	'David Lemmer'	-
129	'Daya'	Mr.. V.N. Palekar & Co., Bombay, India, 1966
130	'Deep Cherry'	AHSI, Kolkata, 1986
131	'Delta Dawn'	New Introduction
132	'Dogstar'	S. Percy-Lancaster, Calcutta, India (1935-40)
133	'Dona Rosita Delight'	Mrs. Lolita Lazarocuruz, Malabon, Rizal, Phillippines, 1962.
134	'Don Fernando'	New Introduction
135	'Don Mario'	New introduction
136	'Donyo'	-
137	'Doorthy Jivara Jadasa'	Madras, India
138	'Double Deligh't	New Introduction
139	'Doubloom'	Philippines, R.E. Holttum, 1970
140	'Dr. B.P. Pal'	S.N. Zadoo & T.N. Khoshoo, NBRI, Lucknow, India, 1969
141	'Dr. H.B. Singh' (Syn. 'Bambino Krishna')	I.I.H.R, Bangalore, 1979
142	'Dr. P.V. Sane'	R. K. Roy, NBRI, Lucknow, India, 2011
143	'Dr. Rao'	-
144	'Dr. R.R. Pal'	Dr. B.P. Pal, New Delhi, India,1959
145	'Dream'	Dr. P.S. Swaminathan Nursery, Madras, India
146	'Dulcia Dabron'	Kenya, Africa
147	'Durga's Delight'	M/S. Chandra Nursery, Sikkim, India
148	'Dwarf Gem'	S. Percy Lancaster, Lucknow, India, 1959
149	'Easter Parade'	Mike Fascell, Coral Way Gardens, Florida, U.S.A
150	'Eclipse'	S. Percy Lancaster Calcutta, India (1935-40)
151	'Equador Pink'	Mrs. J.S. Rover, Trinidad, 1920
152	'Eisha'	Verna Nagpal, Bombay, India, 1984.
153	'Elizabeth'	Introduction from Africa at LBG, India, by Mr. H.P. Greensmith, 1961
154	'Elizabeth Angus'	Kenya, Africa
155	'Elizabeth Doxey' (Syn. 'Jamaica White', 'Medonna', 'White Cloud')	Introduction at Florida, USA by Norman Reasoner, 1926
156	'Ena Heyneker'	Rodney x Jonklaas Jaela, Ceylon
157	'Enid Lancaster' (Syn. 'Golden Queen')	A. Percy-Lancaster, Delhi, India, 1930
158	'Enid Walker'	A. Percy-Lancaster, Lucknow, India

159	'Espinosa'	-
160	'Ethiraj'	AHSI, Madras, India
161	'Evelyn Jeff'	Jamaica, 1938
162	'Evita'	New Introduction
163	'Excelsior'	W.F. Turley, Queensland, Australia
164	'Excelsior Red'	New Introduction
165	'Exquisite'	S. Percy-Lancaster Calcutta, India (1935-40)
166	'Fantasy'	B. Rama Rao, Madras, India'
167	'Fantasy Red'	New Introduction
168	'Feathery Fantasy'	Introduced from Thailand
169	'Feathery Fantasy Bi-colour'	Introduced from Thailand
170	'Fiesta' (Hot Pink)	New Introduction
171	'Filomen'	-
172	'Fitina' (Purple)	-
173	'Flame'	M/S. Soundarya nurseru, Madras, India.
174	'Flame Red'	-
175	'Flamingo'	-
176	'Floribunda'	Introduction at LBG, Bangalore, India by Mr. H.P. Greensmith, 1961
177	'Formosa'	Brazil, introduced in India by Thomas Royer at LBG, Bangalore, India.
178	'Gagarin'	Lalbagh Botanic Gardens, Bangalore, India (1960)'
179	'Gangamma'	Sh. Y. Anivenkatappa, Nurseryman, Bangalore, 1974
180	'Gangaswamy'	Lalbagh Botanical Gardens, Bangalore, 1974
181	'Garnet Glory'	Rodney Jonklaas, Jaela, Sri Lanka
182	'Gem'	B.S. Nirody, Indian Institute of Science, Bangalore, India
183	'Gillian Greensmith'	Introduction from Kenya, Africa into Ceylon
184	'Glabra'	Brazil, 1860'
185	'Glabra Magnifica Trilli'	-
186	Glabra Magnifica	-
187	'Glabra Sanderiana'	-
188	'Glabra Variegata'	In India from Kenya, Africa in 1961 through Mr. H.P. Greensmith.
189	'Glady's Heburn'	Introduction from Kenya, Africa in 1961 through Mr. H.P. Greensmith
190	'Gloriosus'	M/S. Soundarya Nursery, Madras, India
191	'Gloucester Royal'	New Introduction
192	'Glowing Flame'	New Introduction
193	'Godrej Cherry Blossom' (Syn. 'Godrej Centenary')	K. V. Krishna Rao, 1997
194	'Gokul'	M/S SoundaryaNursery, Madras, India
195	'Golden Double'	-
196	'Golden Giant'	-

197	'Golden Glow'	Origin - Cuba, Introduction from Africa at LBG, India in 1961
198	'Golden Ice'	New Introduction
199	'Golden Jackpot'	New Introduction
200	'Golden Queen'	A. Percy-Lancaster Delhi, India, 1948'
201	'Golden Summer'	-
202	'Golden Tango'	-
203	'Gold Rush'	-
204	'Gopal'	M/S K.S. Gopalaswamienger Son, Bangalore, India, 1935
205	'Gulaby'	M/S Soundarya Nursery, Madras, India
206	'Happiness'	M/S. K.S. Gopalaswamienger Son, Bangalore, India
207	'Harlequin'	New Introduction
208	'Harissi'	Reported from Srilanka by Rodney Jonklass
209	'Hawaiian Beauty'	NBRI, 1990
210	'Hawaiian Gold'	New Introduction
211	'Hawaiian Pink'	New Introduction
212	'Hawaiian White'	New Introduction
213	'Helen Coppinger'	Royal Palm Nurseries (Reasioner Brothers) Florida, U.S.A.
214		
215	'Helen Johnson'	'J.E. Hendry, Everglades Nursery, Florida, U.S.A., 1927
216	'Hensbergii'	Pretoria, South Africa
217	'Hiawatha'	S. Percy-Lancaster,1959, Lucknow
218	'Himani' / 'Miss Alice'	Thailand
219	'Hugh Evans' (Syn. Blondie)	Golby Eric V., 1970
220	'Inca Gold'	New Introduction
221	'Intermedia'	S. Percy-Lancaster Calcutta, India (1935-40).
222	'International Red'	-
223	'Isabel Greensmith'	Introduced from Kenya, Africa in India at LBG by H.P. Greensmith, 1961
224	'Isla Morada'	-
225	'Jamaica Pink'	Bert Kraft, Fort Lauderdale, Florida, U.S.A., 1965
226	'Jamaica White'(Syn. Elizabeth Doxey', 'Madonna', 'White Cloud').	Florida , USA
227	'Jamburi'	Kenya, Africa.
228	'James Walker'	Agri-Horticultural Society, Madras, India
229	'Jane Snook' (Lilac Ting)	-
230	'Jasper'	Florida, USA from Cuba.
231	'Jasper Rose'	AHSI, Madras, India
232	'Java White'	-
233	'Jawaharlal Nehru'	I.I.H.R. Bangalore, 1975
234	'Jaya'	-

235	'Jayalakshmi'	M/S K.S. Gopalaswarnienger Son, Bangalore India, 1948
236	'Jayalakshmi Variegata'	Abraham and Desai, BARC, 1977
237	'Jennifer Fernie'	Introduced in India by Mr. H.P. Greensmith, from Kenya, Africa, 1961
238	Jennifer Nagpal	Verna Nagpal, Bombay, India, 1984
239	'Joe de Lovera'	-
240	'John Lettin'	Kenya, Africa
241	'Juanita Hatten'	-
242	'Jubilee'	S. Percy-Lancaster, Calcutta, India (1935-40)
243	Showlady'	-
244	'Kalyani'	M/S K.S. Gopalaswarnienger Son, Bangalore, India (1965).
245	'Kayata'	Introduced from Kenya, Africa by Mr. H.P. Greensmith, India, 1961
246	'Key West Alba'	-
247	'Killie Campbell'	Durban, S. Africa; introduced in India in 1961
248	'Killie Campbell Variegata'	Raised in Nairobi, Kenya, Africa.
249	'Kiriga Bronze'	A.W. Burnley, Kenya, Africa
250	'Klong Fire'	-
251	'Krumbiegal'	M/S K.S. Gopalaswarnienger Son, Bangalore, India,1948
252	'Kuvempu'	Lalbagh Botanic Gardens, Bangalore, India, 1960
253	'L.N. Birla'	AHSI, Kolkata, India, 1962
254	'Lady Bird'	New Introduction
255	'Lady Higgins'	Raised in Trinidad. Imported from Jamaica to Kenya by H.A. Delap
256	'Lady Watts'	-
257	'La Jolla'	-
258	'La Joya'	-
259	'Lady Hope'	B. Rama Rao, Madras, India.
260	'Lady Hudson' (Syn. Princess Margaret Rose)	M/s L.R. Russell, England, 1938
261	'Lady Hudson of Ceylon Variegata'	B.M. Desai & V. Abraham, BARC, Bombay, India, 1979
262	'Lady Baring' ('Yellow Glory' / 'Hawaiian Gold')	-
263	'Lady Casimir'	-
264	'Lady Mary Baring'	Kenya, Africa in 1961 through Mr. H.P. Greensmith
265	'Lady Mountbatten'	Soundarya Nursery, Madras & Imperial Nursery, Calcutta, India
266	'Lady Richards'	Introduced in india by LBG, bangalore, India, 1961
267	'Lady Seton James'	W.N. Sands, St. Vincent, Peru
268	'Lady Stubbs'	Raised at Jamaica,1938
269	'Lady Watts'	St. Vincent, Peru, 1919
270	'Lady Wilson	Jamaica,1938
271	'Lafitte' (Red)	-
272	'Lalbagh'	Lalbagh Botanic Gardens, Bangalore, India

273	'Lateritia' (Syn. 'Dar-es-Salaam')	London, 1865
274	'Lateritia Jr.'	Produced in east coast of U.S.A.
275	'Lavender Glory'	Rodney Jonklaas Jaela, Sri Lanka
276	'Laxmi Narayana'	Lalbagh Botanic Gardens, Bangalore, India, 1961
277	'Lazat of Mysore'(Syn. 'Maharaja of Mysore')	Kenya, E. Africa
278	'Lehmanii'	Ecuadore, 1932
279	'Lemmers Pride'	New Introduction
280	'Lemmer Special'	
281	'Lewis'	Lewis Nursery, Military, Trail, Lake Worth, Florida, U.S.A.
282	'Lilac Puff'	-
283	'Lilac Perfection'	-
284	'Lilac Queen'	M/s Gopalaswamiengar son, Bangalore, India, 1956.
285	'Lilacina'	M/S. Soundarya Nursery, Madras, India
286	'Limberlost Beauty'	Grand Canary Islands
287	'Limousine'	-
288	'Lindleyana' (Syn. 'Aurantiaca')	Grand Canary Islands
289	'Lip Stick'	Thailand
290	'Little Node'	Thailand
291	'Lord Willingdon'	S. Percy-Lancaster, Calcutta, India, 1934
292	'Los Banos Beauty'	Reported by J.V. Pancho (1967), College of Agriculture Laguna, Philippines
293	'Los Banos Beauty Variegata'	S.K. Datta, B.K. Banerji & S.C. Sharma, N.B.R.I., Lucknow, India, 1990
294	'Los Banos Variegata 'Jayanti'	Jayanthi and Datta, 2006
295	'Los Banos Variegata 'Silver Margin'	Banerji, NBRI, Lucknow, India, 2002
296	'Louise Wathen' (Syn. 'Orange Glory')	B.S. Nirody, Agri-Horticultural Society, Madras India, 1932
297	'Louise Wathen Intermediate'	Imperial Nursery, Calcutta, India
298	'Louise Wathen Mediopicta'	Royal Agri- Horticultural Society of India, Calcutta, India (1935-40)
299	'Louise Wathen Variegata'	Royal Agri-Horticultural Society of India, Calcutta, India (1935-40)
300	'L.N. Birla'	AHSI, Kolkata
301	'Lucifer Red'	-
302	'Machakos'	Introduce by LBG, by Mr. H.P. Green Smith, Bangalore, India, 1961
303	'Madonna' (Syn. 'Elizabeth Doxey', 'Jamaica White' & 'White Cloud')	M/S Evans and Reeves, California, U.S.A.
304	Magenta Queen (Syn. 'Mr. Butt Magenta', 'Purple King', 'Purple Queen', 'Purple Pink').	B. Rama Rao, Madras, India, 1945.

305	'Magnifica'	Australia , 1893.
306	'Magnifica var. Traillii'	Australia, 1893
307	'Mahara' (Syn. 'Million Dollar, 'Mantila', 'Magic Red', 'Mahara Crimson', 'Manila Red)	Lalbagh Botanic Gardens, Bangalore, India, 1961.
308	'Mahara Beauty'	New Introduction
309	'Mhara Magic'	New Introduction
310	'Mahara Variegata'	S.K. Dutta & B.K Banerji, N.B.R.I, Lucknow, 1991
311	'Mahara Variegata 'Abnormal'	B.K. Banerji, NBRI, 2002
312	'Maharaja of Mysore'	B.S. Nirodi, LBG, India, 1938
313	'Mahatma Gandhi'	K.S. Gopalaswamienger Son, Bangalore, India
314	'Mahatma Gandhi Variegata'	Pratap Nursery, Dehradun, India, 1969
315	'Manila Magic Red'	Edwin A. Meninger
316	'Manohar Chandra'	M/S Chandra Nursery, Sikkim
317	'Manohar Chandra Variegata	S.C. Sharma and K.K. Basario, NBRI, Lucknow, India, 1985
318	'Marandi Gras'	Thailand
319	'Margaret Bacon'	J.E. Hendry, Everglades Nursery, Florida, U.S.A.,1927
320	'Margaret Rose'	Java
321	'Margery Lloyd'	Introduced in LBG, Bangalore, India by Mr. H.P. Greensmith, 1961.
322	'Mariel Fitzpatric'	Introduced in LBG, Bangalore, India by Mr. H.P. Greensmith, 1961
323	'Marietta'	Dr. J.V. Pancho (1967), College of Agriculture, Laguna, Philippines
324	'Marigowda'	Department of Horticulture , Lalbagh, Bangalore, India
325	'Marilyn-Hatten'	New Introduction
326	'Mardi Gras'	New Introduction
327	'Margery Lloyd'	Introduced in LBG, Bangalore, India by Mr. H.P. Greensmith, 1961
328	'Mary Palmer' (Syn. 'Surprise' as patented in U.S.A.)	S. Percy Lancaster, Calcutta, India ,1949
329	'Mary Palmer Enchantment'	New Introduction
330	'Mary Palmer Special'	S.N. Zadoo and T.N. Khoshoo NBRI, Lucknow, India, 1974
331	'Mataji Agnihotri'	-
332	'Maude Chettleburgh'	London in 1900 by Col. Rous, Norwich, England
333	'Maureen Hatten'	-
334	'Mauve Queen'	Soundarya Nursery, Madras, India
335	'Meera' (Syn. Mrs. Butt Scarlet)	A.R. Rangachari Madras, India, 1930
336	Meera Sport	A.R. Rangachari Madras, India, 1930
337	'Miami Pink'	New Introduction
338	'Midget'	S. Percy Lancaster, Calcutta, India (1935-40)
339	'Millari'	Kenya, Africa.
340	'Million Dollar' (Syn. Mahara)	Reported by Dr. J.V. Pancho, 1963
341	'Mini Thai'	Thailand

342	'Minyata'	Kenya, Africa
343	'Miss Alice'	New Introduction
344	'Miss Luzon'	Reported by Dr. J.V. Pancho and Bardenas, Laguna, Philippines, 1959
345	'Miss Manila'	Reported by Dr. J.V. Pancho and Bardenas, Laguna, Philippines, 1959
346	'Miss. Universe' (Syn. 'Aiskrim')	-
347	'Mong Heng'	Named by R.E. Holttum, Singapore, 1940
348	'Moonlight Madonna'	John Popenol, 1961
349	'Mrs. Vera Blakeman'	New Introduction
350	'Mrs. A.T. Stephenson'	Australia.
351	'Mrs. Butt' (Syn. Ruby Crimson Lake)	W.F. Turkey, Queensland, Australia
352	'Mrs. Butt Magenta' (Syn. 'Magenta Queen' or 'Purple Queen')	R.O. Williams, 1923
353	'Mrs. Butt Scarlet '(Syn. 'Meera')	S. Percy Lancaster, Lucknow, 1959
354	'Mrs. Butt Variegata'	-
355	'Mrs. Chico'	Rodney Jonklaas Jaela, Sri Lanka'
356	'Mrs. Deluz Perry'	A. Percy Lancaster, Delhi, India
357	'Mrs. E.W. Bick'	-
358	'Mrs. Eusenia Raja Singh'	W.F. Turley Queensland, Australia
359	'Mrs. Eva Ice Cream'	-
360	'Mrs Eva Purple'	Don Justin Anadetus Raja Singh
361	'Mrs Eva Mauve Variegata'	New Introduction
362	'Mrs. Fraser'	New Introduction
363	'Mrs. H.C. Buck'	B.S. Nirody, Lalbagh Botanic Gardens, Bangalore, India, 1932.
364	'Mrs. Lancaster'	P.S. Swaminathan, Soundarya Nursery, Madras, 1930.
365	'Mrs. Leono'	S. Percy Lancaster, Calcutta, India (1935-40).
366	'Mrs. Mc Clean' (Syn. 'Orange King')	Reported by Dr. J.V. Pancho and Bardenas, Laguna, Philippines, 1959
367	'Mrs. Oliver Perry'	Originated at Trinidad in 1931. Named by R.O. Williams
368	'Mrs. R.B. Carrick'	W.F. Turley, Queensland, Australia
369	'Mudanna'	B. Rama Rao Madras, India
370	'Munivenkatappa'	Sh. Y. Muniven Katappa Nurserymen, Bangalore
371	'Natalii'	Durban, S. Africa; introduced in India at LBG, Bangalore, India, 1961
372	'Nawab Ali Yawar Jung'	Mr. Habibur Rehman, Aligarh India
373	Netta Zukerman	Verna Nagpal Bombay, 1986
374	New Red	Royal Agricultural Horticultural Society of India, Alipore Road, Calcutta, India, 1968
375	'New River'	-
376	'Nigrette'	New Introduction

377	'Nirmal'	G.S. Srivastava, NBRI, 1982
378	'Nirmal Chandra'	Chandra Nursery, Sikkim
379	'No. 2'	Introduced in India from Kenya, Africa at LBG, Bangalore, by Mr. H.P. Greensmith
380	'Nonya'	Australian Bambino variety
381	'Odisee'	P. Das, O.U.A.T. Bhubaneswar, 1977
382	'Ole'	-
383	'Oo-La-La'	New Introduction
384	'Orange Chilli'	New Introduction, Thailand
385	'Orange Glory' (Syn. 'Louis Wathen')	S. Percy Lancaster Calcutta, India (1935-40)
386	'Orange Ice'	Thailand
387	'Orange King'	Evans & Reeves, Los Angles, USA.
388	'Orange Queen'	Evans & Reeves Los Angels, U.S.A
389	'Orange Sceptre'	M/S. K.S. Gopalaswamienger son, Banglore, India, 1954
390	'Orange Sunset'	-
391	'Padmi'	M/S. K.S. Gopalaswamienger son, Banglore, India, 1954
392	'Pagoda Pink'	-
393	'Palekar'	Palekar, Bombay, India
394	'Pallavi'	B.K. Banerji & S.K. Datta N.B.R.I, Lucknow, 1987
395	'Panama Pink'	Named by Egbert Reasoner, Florida, U.S.A., 1901
396	'Partha' (Syn. 'Indian Flame' in Africa)	M/S KS Gopalaswamienger son, Banglore, India, 1954
397	'Parthasarathy'	S.C. Sharma, N.B.R.I., Lucknow , 1974
398	'Pearl'	Boseck & Co., Calcutta, India, 1970-71
399	'Penang' (Syn. 'Coker')	Named by R.E. Holttum at Singapore in 1938
400	'Penelope'	Jan Iredell, Australia, 1986
401	'Perfect'	Rodney Jonklaas, Jaela, Sri Lanka
402	'Perfection'	M/S. Soundarya Nursery, Madras India
403	'Philips No. 1'	Introduced in India by Mr. H.P. Greensmith, at LBG, 1961
404	'Phoenix'	Jan Iredell, Australia, 1986
405	'Pigeon Blood'	J.V. Pancho and A.E. Bardenas, Philippines, 1959
406	'Picta Aurea'	Thailand
407	'Pink Beauty'	M/S Soundarya Nursery, Madras India
408	'Pink Clusters'	-
409	'Pink Champagne'	-
410	'Pink Delight'	New Introduction
411	'Pink Pearl'	New Introduction
412	'Pisil'	New Introduction
413	'Pixie'	M/S. Soundarya Nursery, Madras, India.
414	'Pixie Pink'	New introduction
415	'Pixie Orange'	New introduction
416	'Pixie Queen'	New introduction

417	'Pixie Variegata'	B.K.Banerji, NBRI, Lucknow
418	'Plum Crazy'	New Introduction
419	'Poultoni'	Durban, S. Africa; introduced in India, at LBG, by Mr. H.P. Greensmith, 1961
420	'Poultoni Special'	Durban, S. Africa; introduced in India, at LBG, by Mr. H.P. Greensmith, 1961
421	'Poultoni Vareigata'	Abraham and Desai, BARC, Trombay, 1986
422	'Pradhan's Pink'	Chandra Nursery, Sikkim
423	'Pradhan's Profusion'	Chandra Nursery, Sikkim
424	'Praetorius' (Syn. 'Mrs. cClean')	Java
425	'Preeti'	V.N. Palekar, Palekar & Co., Bombay, 1970
426	'President'	Reported by S.C. Sharma N.B.R.I., Lucknow, 1996
427	'President Roosevelt'	Soundarya Nursery, Madras, India.
428	'Pride of Rhodesia'	Scott, Banket, Rhodesia.
429	'Pride of Singapore'	R.E. Holttum, 1943
430	'Princess Elizabeth' (Syn. 'Rosa Catalina')	Grand Canary Islands
431	'Princess Margaret Rose' (Syn. 'Lady Hudson')	Agri Horticultural Society, Madras, India, 1935
432	'Princess Thai'	-
433	'Profusion'	S. Percy-Lancaster, Calcutta, India (1935-40)
434	'Purple Gem'	M/S. Soundarya Nursery, Madras, India
435	'Purple King' (Syn. 'Magenta Queen')	Rama Rao, Madras, India 1948
436	'Purple Prince'	Rama Rao, Madras, India, 1948
437	'Purple Queen' (Syn. 'Magenta Queen')	Rama Rao, Madras, India, 1948.
438	'Purple Robe'	S. Percy Lancaster, Calcutta, India (1935-40)
439	'Purple Star'	M/S Soundarya Nursery, Madras, India.
440	'Purple Wonder'	I.I.H.R, Bangalore, 1979
441	'Queen Violet'	New Introduction
442	'Rainbow'	J.V. Pancho and A.E. Bardenas, Laguna Philippines, 1959
443	'Raman'	Lalbagh Botanic Gardens, Bangalore, India
444	'Ranee'	M/S Soundarya Nursery, Madras, India
445	'Rao'	S. Narsingarao, Central Food Technological Research Institute, Mysore, India, 1954
446	'Raspberry Ice'	New Introduction
447	'Raspa Raspbery'	New Introduction
448	'Ratna'	M/S Chandra Nursery, Sikkim
449	'Red Glory'	M/S KS Gopalaswamienger son, Banglore, India, 1954
450	'Red Glory Improved'	M/S KS Gopalaswamienger son, Banglore, India, 1954.
451	'Red Lantern'	-
452	'Red September'	-

453	'Red Sleeping Beauty'	New Introduction
454	'Red Tiger'	-
455	'Red Triangle'	-
456	'Red Zed'	New Introduction
457	'Refulgens'	Brazil; introduced in India at LBG, by Mr. H.P. Greensmith, 1961
458	'Refulgens Variegata'	M/s KS Gopalaswamienger son, Banglore, India, 1954
459	'Rhodamine'	Kenya, Africa; introduced in India at LBG by Mr. H.P. Greensmith, 1961
460	'Rodney Jonklass'	Mrs. Heyenkar, Jonklass
461	'Rosa Catalina' (Syn. 'Princess Elizabeth')	Introduced to England from Grand Canary, Islands, 1909
462	'Rosa Multiflora'	M/S Soundarya Nursery, Madras, India
463	'Rosea'	New Introduction
464	'Rosa Catalina'	-
465	'Rosalane'	R.E. Holttum, 1970
466	'Rose Opal'	H. Grahame Bell, Kenya, Africa.
467	'Rose Queen'	M/S Soundarya Nursery, Madras, India.
468	'Rosea'	Introduced into Florida (U.S.A.) by Egbert Reasoner in 1865 from Hope Gardens, Jamaica
469	'Rosea Fuchsia'	Agri-Horticultural Society, Madras, India
470	'Rosenka'	Kenya, Africa
471	'Roseville's Delight' (Syn. 'Dona Rosita Delight', 'Doubloon')	Mrs. Lolita, Malabon Rizal, Philippines, 1962
472	'Royal Bengal'	New Introduction
473	'Royal Dauphine'	Thailand
474	'Royal Purple'	Thailand
475	'Ruarka'	Introduced in LBG, India by Mr. H.P. Greensmith, 1961from Kenya, S. Africa
476	'Rubra'	-
477	'Ruby'	S. Percy-Lancaster, Lucknow, India, 1959
478	'Sachidananda	M. Sachidananda, Yedur Botanic Gardens, Bangalore, India, 1956
479	'San Diego Red'	Evan & Reeves, Los Angels, U.S.A
480	'San Migud'	-
481	'Sao Paulo'	-
482	'Sanderiana'	M/S. F. Sanders & Co., St. Albans
483	'Saitri'	-
484	'Savitee'	New Introduction
485	'Scarlet Glory'	'M/S. K.S. Gopalaswamienger son, Banglore, India, 1954
486	'Scarlet O'Hara'	New Introduction
487	'Scarlet Queen'	Mr. Tomlinson, Eastern Bengal Railway, India 1920
488	'Scarlet Queen Variegata'	S. Percy-Lancaster, Calcutta, India (1935-40)
489	'Sensation'	M/S. Soundarya Nursery, Madras, India

490	'Sharma'	M.D. Sharma, Floriculturist, Lalbagh Botanic Gardens, Bangalore, India 1961
491	'Sharon Wesley'	-
492	'Sholay'	I.I.H.R., Bangalore 1977
493	'Shubhra'	S.C. Sharma, N.B.R.I., Lucknow, India, 1965.
494	'Shweta'	S.C. Sharma, N.B.R.I., Lucknow, India, 1979.
495	'Silhouette'	-
496	'Silver Top'	New Introduction
497	'SilverLine'	BARC, Mumbai, India
498	'Simon Anselm'	-
499	'Singapore Beauty'	Verna Nagpal Bombay, 1986
500	'Singapore Dark Red'	R.E. Holttum, Singapore
501	'Singapore Pink'	New Introduction
502	'Singapore White'	New Introduction
503	'Sir Roy' (Magenta)	Scott, Banket, Rhodesia.
504	'Smoky'	Rodney Jonklaas, Ceylon
505	Snow-Cap	-
506	'Snow Purple'	New Introduction
507	'Snow Queen'	Introduced in India by late Dr. Homi Bhabha, Bombay
508	'Snow White' (Syn. 'Millionaire')	Introduced in India from Jamaica, Named by R.E. Holttum
509	'Sonnet'	B.P. Pal, New Delhi, India
510	Soundarya	M/S Soundarya Nursery, Madras, India
511	'Sova'	-
512	'Speciosa' (Syn. 'spectabilis var. Speciosa')	Described in 1849, figured from plants grown in Britain, 1854
513	'Spectabilis'	Peru, 1835
514	'Spectabilis var. Variegated'	S. Percy-Lancaster (1959) Lucknow, India
515	'Spitfire'	S. Percy-Lancaster, Calcutta, India (1935-40)
516	'Splendens'	Introduced in India by LBG from Kenya, S. Africa by Mr. H.P. Greensmith, 1961
517	'Splendens Improved'	Central Africa
518	'Spring Festival'	B.P. Pal, New Delhi, India, 1959
519	'Sprinkle Gold Safflower'	
520	'Srimati Biezli'	M/s Chandra Nursery, Sikkim
521	'Sri Durga'	-
522	'Srinivasa'	M/S K.S. Seedling of Gopalaswamienger Son, Bangalore, India, 1965
523	'Stanza'	B.P. Pal, New Delhi, India, 1964
524	'Star Mauve'	AHSI, Kolkata, India.
525	'Suicheng 25-2'	New Introduction
526	'Sumali'	-

527	'Summer Breeze' (Orange)	New Introduction
528	'Summer Breeze' (Red)	New Introduction
529	'Summer Breeze' (Pink)	New Introduction
530	'Summer Time'	B.P. Pal, New Delhi, India, 1959
531	'Sundown Orange'	-
532	'Sundri'	-
533	'Sunvillea Cream'	-
534	'Sunvillea Pink'	-
535	'Sunvillea Rose'	-
536	'Superba'	M/S Soundarya Nursery, Madras India
537	'Surekha'	S. C. Sharma, N.B.R.I. , Lucknow, 1981
538	'Surprise' (Syn. 'Mary Palmer')	R.E. Holttum in E. A. Menninger's book
539	'Surprise Bouquet'	New Introduction
540	'Susan Hendry'	J.E. Hendry, Everglades Nursery, Florida, U.S.A., 1927
541	'Suverna'	BARC, Mumbai, India
542	'Sweetheart'	M/S Soundarya Nursery, Madras India
543	'Sydney'	Introduced in India at LBG by Mr. H.P. Greensmith, 1961
544	'Sylvia Delap'	Mrs. D.A. Delap, Kenya, Africa
545	'Tahitian Dawn'	-
546	'Tango'	New Introduction
547	'Temple Fire' (Syn. 'Closeburn')	Evans & Reeves, Los Angeles, U.S.A
548	'Tequila Sunrise'	-
549	'Tetra Mrs. McClean'	S.N. Zadoo & T.N. Khoshoo, N.B.R.I., Lucknow, India, 1969
550	'Texas Dawn'	Evans & Reeves, Los Angeles, U.S.A.
551	'Thai Cherry'	Thailand
552	'Theresa Jacob'	-
553	'Thimma'	Lalbagh Botanic Gardens, Bangalore, India
554	'Thimma Sport'	KSG, Bangalore , India
555	'Thomasii' (Syn. Rosea)	Mr. Thomas, Brisbane, Queensland, Australia, 1905
556	'Tomat Red'	W.F. Turley, Queensland, Austrtalia
557	'Torch Glow'	-
558	'Traillii'	New Introduction
559	'Treasure'	New Introduction
560	'Tricolor'	
561	'Trinidad'	Reported by M.H. Marigowda, 1960 Lalbagh Botanic Gardens, Bangalore, India
562	'Tropical Bouquet'	-
563	'Tropical Rainbow'	-
564	'Turley's Special'	W.F. Turley, Queensland, Australia , 1930
565	'Twilight Delight'	-
566	'Tyrian Rose'	Agri-Horticultural Soc., Madras, India
567	'Udai Chandra'	Chandra Nursery, Sikkim, India

568	'Usha'	I.I.H.R., Bangalore, India
569	'Vellayani'	Reported by M.H. Marigowda (1960) Lalbagh Botanic Gardens, Bangalore India
570	'Vera Deep Purple'	-
571	'Verna Nagpal'	Verna Nagpal, Bombay, 1984
572	'Versicolour'	Soundarya Nursery, Madras, India
573	Vesuvius	'S. Percy-Lancaster, Calcutta, India (1935-40)
574	'Vicky'	-
575	'Vijaya' (Syn. 'Mahatma Gandhi')	Soundarya Nursery, Madras India
576	'Violet	New Introduction
577	'Vishakha'	B. Singh, R.S. Malik, L.P. Yadav and B. Choudhury, I.A.R.I., New Delhi
578	'Vishakha Variegata'	-
579	'Vittal Variegata'	Seedling selection, Madras, India
580	'Wajid Ali Shah'	S.N. Zadoo and T.N. Khoshoo, N.B.R.I. Lucknow, 1977
581	'Wall Flower'	Hazlewood Bros., New South Wales, Australia
582	'White Cloud' (Syn. 'Elizabeth Madona', 'Jamica White')	Florida , USA
583	'White Stripe'	-
584	Winsome	-
585	'Yani's Delight' (Pink)	S. Percy-Lancaster, Calcutta India (1935-40).
586	'Yuehong 85-1'	-
587	'Yellow Lantern'	-
588	'Yellow Queen'	Reported by M.H. Marigowda, Lalbagh Botanic Gardens, Bangalore, India
589	'Yuyu' (Blue)	-
590	'Zakir Hussain'	Lalbagh Botanic Gardens, Bangalore, 1963
591	'Zakiriana'	-
592	'Zakiriana Variegata'	AHSI, Kolkata
593	'Zinna Barat'	S.C. Sharma, N.B.R.I., Lucknow, 1996
594	'Zulu Queen'	-

B. glabra Mrs. 'Eva Variegata'

B. peruviana 'Mary Palmer Special'

B. peruviana 'Palekar'

B. peruviana 'President Roosevelt'

B. spectabilis 'Aida'

B. spectabilis 'Splendens'

B. buttiana 'Los Banos Variegata'

B. buttiana 'Mahara'

B. buttiana 'Roseville's Delight'

B. buttina 'Cherry Blossom'

B. glabra 'Glabra'

B. glabra 'H.B. Singh

*****

4

Characterization of Varieties - Morphological and Floral

Characterization is a study, evaluation and documentation of various morphological (vegetative) and floral characters of ornamentals or any other plant species. In this method, horto-taxonomical studies are done and details of vegetative and floral characters are documented for the purpose of correct identification and establishing identity of the new species/ varieties.

There are standard methods and approaches for characterization of new varieties. Depending upon the type of plant *viz.,* trees, shrubs, herbs, bulbous and rhizomatous, characters to be studied are decided.

In case of Bougainvillea, following morphological characters are mainly considered for characterization.

Vegetative Characters - Growth habit, plant height, thorns (size, shape), leaves (leaf colour -young and mature, texture, size, shape) (Fig. 1A, B).

Floral Characters - Flowering habit (profuse, sparse, medium, season), inflorescence, bracts (colour at flowering, change of colour from young to old bracts, size, margin, base, shape, persistent or non-persistent after flowering, flower tube and star (Fig. 1C).

Fig. 1. *Bougainvillea*: **(A) Types of spines, (B) Leaf characters (C) Bracts with flower**

Significance of Characterization

Study of morphological characters is most important for identification purpose and establishing identity of the variety. Both vegetative and floral characters are studied, recorded and interpreted in a scientific and standard manner. Each and every variety has got certain diagnostic characters which differ from the others. Morphological characterization identify those diagnostic characters which actually serve as tool for identification. Therefore, studies on morphological characters have enormous importance both taxonomically and horticulturally.

There are eighteen species and hundreds of varieties of Bougainvillea available all over the world. Each species has got some diagnostic key characters which are basically morphological traits consisting of both vegetative and floral (bracts) parameters. Similarly, varieties are also differentiated on the basis of morphological characters. Visual recognition by the bract colour is the easy and commonly followed method. Sometimes, overlapping bract colour and other vegetative characters closely resembled resulting confusion and wrong identity of the varieties.

Considering the above and to eliminate the error of identification, there are some identified morphological traits which have been furnished hereunder in Table – 4.1. On the basis of these characters, differentiation of the varieties and their identification is possible.

Table – 4.1: Morphological characters (vegetative & floral) and description of the plant parts

S. No.	Morphological Characters	Details
1	Growth habit	Upright, Semi-upright, Spreading, Drooping, Climbing.
2	Colour of young Shoot	Light green, Medium-green, Reddish-green , Reddish.
3	Length of internodes	Short, Medium, Long.
4	Thorns	Absent, Present.
5	Density of thorns	Sparse, Medium, Dense.
6	Length of thorn	Short, Medium, Long.
7	Curvature of thorn	Straight, Slightly curved, Fully curved.
8	Strength of thorn	Weak, Medium, Strong.
9	Length of leaf blade	Short, Medium, Long.
10	Width of leaf blade	Narrow, Medium, Broad.
11	Shape of leaf blade	Lanceolate, Medium ovate, Broadly ovate, Elliptic, Circular.
12	Apex shape of leaf blade	Acuminate, Acute, Obtuse.
13	Base shape of leaf blade	Attenuate, Acute ,Obtuse.
14	Colour of young leaf	Light green, Medium-green, Reddish-green, Reddish.
15	Main colour of leaf	Yellowish-white, Yellow, Yellowish-green, Light green, Medium green, Dark green, Very dark green, Grey-green.

16	Secondary colour of leaf	None, White, Yellowish-white, Yellow, Light green, Medium-green, Dark green, Very dark green, Grey-green.
17	Distribution of secondary colour on leaf	Absent, Narrow-marginal, Broad-marginal, Around midrib, Speckled, Irregular.
18	Tertiary colour of leaf	None, White, Yellowish white, Yellow, Light green, Medium green, Dark green, Very dark green, Grey green.
19	Undulation of margin	Absent or weak, Medium, Strong.
20	Texture of leaf blade	Glabrous, Hairy, Slightly Hairy, Tomentose.
21	Number of leaf on primary branch	Sparse, Medium, Dense.
22	Persistence of leaf blade	Persistent, Non Persistent.
23	Length of petiole	Short, Medium, Long.
24	Attitude of petiole	Upward, Horizontal, Downward.
25	Length of inflorescence	Short, Medium, Long.
26	Peduncle length	Short, Medium, Long.
27	Arrangement of bract clusters	Terminal, Axillary, Axillary and Terminal.
28	Number of bract clusters	Few, Medium, Many.
29	Density of bract clusters	Sparse, Medium and Dense.
30	Presence of flowers	Absent and Present.
31	Type of bract	Single, Multiple and Double.
32	Length of bract	Short, Medium, Long.
33	Width of bract	Narrow, Medium, Broad.
34	Shape of bract	Narrowly Ovate, Medium Ovate, Broadly Ovate, Circular.
35	Reflection of bract	Reflexed, Normal/ Straight.
36	Tip shape of bract	Acute, Obtuse.
37	Base shape of bract	Acute, Obtuse, Cordate.
38	Persistence of bract	Persistent, Non-Persistent.
39	Star	Prominent, Non-Prominent.
40	Colour of star	White , Creamy, Greenish-yellow, yellow, Red and Pink.
41	Diameter of star	Short, Medium, Broad .
42	Colour of floral tube	Green, Orange, Magenta, Red and Purple.
43	Shape of floral tube	Slender with little constriction in the middle and swollen at base.
44	Stamen position	Inserted and Exerted.
45	Main colour of small young Bract	White, Greenish-White, Yellow, Orange, Magenta, Pink, Red, Mauve and Purple.
46	Main colour of young bract (Calyx lobe/ Star not open)	White, Yellow, Orange, Magenta, Pink, Red, Mauve and Purple.
47	Main colour young bract (Star open)	White, Yellow, Orange, Magenta, Pink, Red, Mauve and Purple.
48	Secondary colour of young bract (Calyx lobe/ Star open)	White, Yellow, Orange, Magenta, Pink, Red, Mauve and Purple.
49	Tertiary colour of young bract (Calyx lobe/ Star open)	White, Yellow, Orange, Magenta, Pink, Red, Mauve and Purple.
50	Main colour of bract (Calyx lobe/ Star wilted / fading)	White, Yellow, Orange, Magenta, Pink, Red, Mauve and Purple.

Table – 4.2: Schematic diagram of the various morphological characters.

Characters

GROWTH HABIT

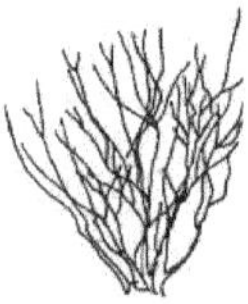 Upright	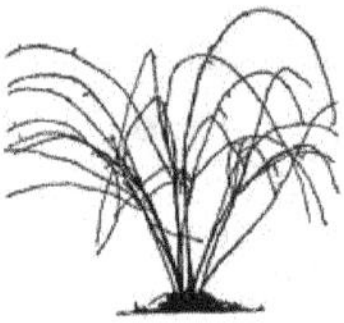 Semi-upright	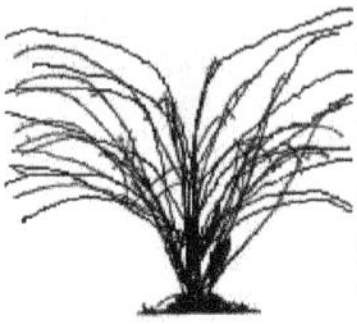 Spreading	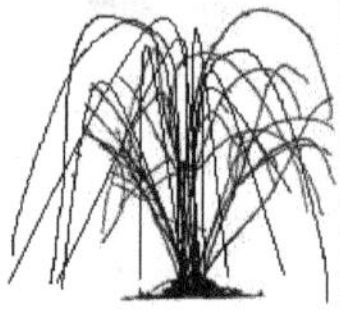 Drooping	 Climbing

LENGTH OF INTERNODES

 Short	 Medium	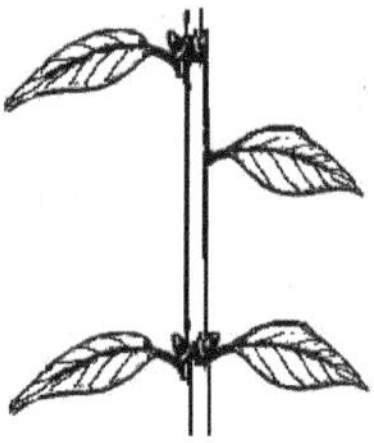 Long

THORNS

 Absent	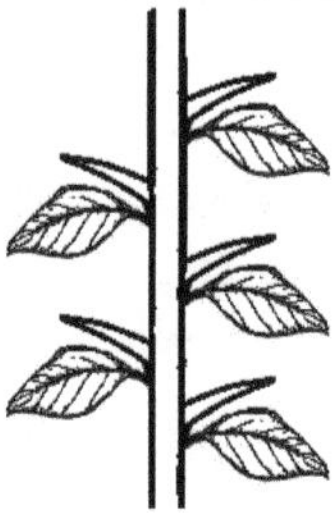 Present

CURVATURE OF THORN

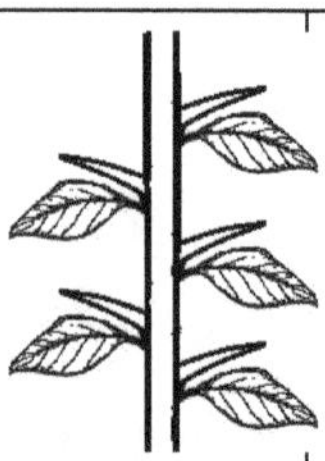 Straight	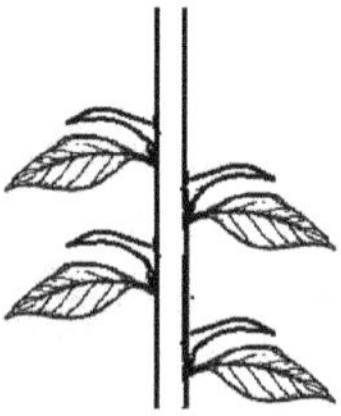 Slightly curved	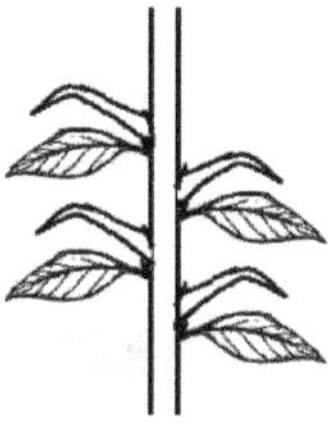 Fully curved

SHAPE OF LEAF BLADE				
Lanceolate	Medium ovate	Broadly ovate	Elliptic	Circular

APEX SHAPE OF LEAF BLADE

Acuminate	Acute	Obtuse

BASE SHAPE OF LEAF BLADE

Attenuate	Acute	Obtuse	Cuneate

DISRIBUTION OF SECONDARY COLOUR ON LEAVES

Absent	Narrow-marginal	Broad -marginal	Around midrib	Speckled	Irregular

UNDULATION OF MARGIN

Absent or weak	Medium	Strong

ATTITUDE OF PETIOLE		
Upward	Horizontal	Downward

PEDUNCLE LENGTH		
Short	Medium	Long

ARRANGEMENT OF BRACT CLUSTERS ON INFLORESCENCE		
Terminal	Axillary	Axillary and Terminal

NUMBER OF BRACT CLUSTERS ON INFLORESCENCE

Few	Medium	More

DENSITY OF BRACT CLUSTERS ON INFLORESCENCE

Sparse	Medium	Dense

PRESENCE OF FLOWERS

Absent	Present

TYPE OF BRACT

Single	Multiple

SHAPE OF BRACT			
Narrowly Ovate	Medium Ovate	Broadly Ovate	Circular

REFLECTION OF BRACT	
Reflexed	Normal/ Straight

TIP SHAPE OF BRACT	
Acute	Obtuse

BRACT: SHAPE AT BASE		
Acute	Obtuse	Cordate

Star: Shape	
Prominent	Non-Prominent
FLORAL TUBE: SHAPE	
Slender with little constriction in the middle	Swollen at base

Note: These diagnostic morphological characters are as per standard international guidelines / parameters following The International Union for the Protection of New Varieties of Plants (UPOV) and guidelines developed by CSIR-NBRI for Protection of Plant Varieties and Farmers' Right Authority, Govt. of India (PPV&FRA in 2014).

*****

5

Genetic Diversity Studies by Molecular Analysis

Morphological characters are basically studied for identification as well as distinguishing one variety from another. It has been found that, in some cases, vegetative and floral characters are overlapping or have minor differences. Therefore, it is very difficult to differentiate varieties on the basis of morphological characters alone, particularly when the availability of varieties is large having wide genetic diversity. All these problems have led to lot of confusion in the identification of varieties (MacDaniels, 1981) and the establishing genetic relationship between parents and their hybrids.

In order to eliminate the above confusion, molecular characterization is very helpful. Moreover, it distinguishes varieties and determines genetic diversity besides studying their relationship. This improves the identification of Bougainvillea significantly and the knowledge on germplasm. In recent years, several molecular characterization techniques are available. Among these, RAPD (Random Amplified Polymorphic DNA) is most commonly used for the identification of varieties due to its simplicity, rapidity and requirement of only a small quantity of DNA to generate numerous polymorphisms (Wight *et. al.* 1993; Cheng *et. al.* 1997). The RAPD assay has been successfully used for studying genetic diversity of many crop species such as Rose (Debener ;. 1996, Martin 2001), *Chrysanthemum* (Sheng *et al..* 2000), *Amaranthus* (Faseela and Joseph 2007) etc. In addition, there are several other methods also very useful for the molecular studies like ISSR (Inter Simple Sequence Repeats) and DAMD (Direct Amplification of Minisatellites) markers. These co-dominant markers are much modified and useful than RAPD. Now *Bougainvillea* is authentically characterized by studying the genetic stock for future utilization for the development of new varieties.

Review of the Work Done

Genetic diversity and relationship among Bougainvillea cultivars at intra and inter-specific level by using RAPD analysis was done by Chatterjee *et.al.* (2007). Fifty random decamer primers were screened and ten were selected for final RAPD analysis. Each primer produced a unique set of amplification products ranging in size from 300-2500 bp. The number of bands for each primer varied from 10 in P2 to 18 in P10. These ten primers used in this analysis yielded 167 scorable bands with an average of 11.3 bands per primer. Of the 167 fragments scored from these primers, 26 were monomorphic and 141 were polymorphic (84.4%). The resulting dendrogram divided the varieties into two main clusters. The first contained 37 varieties and sub-divided into two clusters in which Sub-cluster I had *B. buttiana* and Sub-cluster II had *B. spectabilis* and *B. peruviana* groups. The second major cluster contained 55 varieties which contained species of *B. glabra* group and varieties whose origin were not well recorded. The generated similarity matrix showed that the genetic diversity within the tested genotypes was high. The finding of this work may be useful for better management, identification of accession and also in avoiding duplication or mislabelling of the genotypes studied.

Hammad (2009) studied on the determining the genetic relationships among the most important *Bougainvillea* varieties grown popularly in the gardens in Egypt using Random Amplified Polymorphic DNA (RAPD) technology and isozyme patterns. The main purpose of the study was determining the genetic relationship among the most important Bougainvillea cultivars. Ten random decamer primers were screened and five were selected for final RAPD analysis. Each primer produced a unique set of amplification products ranging in size. Isozyme patterns peroxidase, esterase, poly phenyl-oxidase (PPO) and alcohol-dehydrogenase (ADH) were studied in five cultivars of Bougainvillea having different bract colours. Isozymes revealed higher polymorphism among Bougainvillea cultivars with all enzymes used. The cluster analysis based on isozymes data showed that the entire cultivars fell in one group but the purple and red is very close to each other and orange cultivar was the different.

The molecular studies on 21 varieties of *Bougainvillea* comprising of 9 hybrids and their parents through RAPD marker was done by Srivastava *et.al.* (2009). The UPGMA base dendrogram divided 21 varieties into two major groups with similar coefficient ranging from 0.51 to 0.942. Group A had three varieties namely Trinidad, Formosa and Dr. H.B. Singh in which Dr. H.B. Singh was confirmed as a hybrid of other two varieties. Group B was sub-divided into 8 clusters. The parentages of 7 out of 8 hybrids have been confirmed based on clusters. The study concluded that the RAPD technique is suitable for confirmation of parent-hybrid relationship.

Studies on genetic diversity and phylogenetic relationship of ornamental germpalsm resources in Bougainvillea were carried out by Yan and Jiang (2012) in Huaqiao University, China. They studied genetic diversities and phylogenetic relationship of 48 Bougainvillea cultivars by RAPD, ISSR and SRAP molecular markers. Seven RAPD random primers amplified a total of 97 DNA loci, percentage of polymorphic loci was 100%. PIC, observed number of allies, effective number of alleles, Nie's gene diversity and Shannon's information index were respectively 0.890, 2.0000, 1.4595, 0.2830, 0.4340. Eleven ISSR primers amplified a total of 140 DNA loci, percentage of polymorphic loci was 100%. PIC, observed number of allies, effective number of alleles, Nie's gene diversity and Shannon's information index were respectively 0.835, 2.0000, 1.4189, 0.2579, 0.4040. Twenty five pairs of SRAP primers amplified a total of 773 DNA loci, percentage of polymorphic loci was 97.02 %. PIC, observed number of allies, effective number of alleles, Nie's gene diversity and Shannon's information index were respectively 0.9463, 1.9702, 1.4721, 0.2818, 0.4314. A high genetic diversity level was revealed by the parameters studied.

Molecular characterization and cultivar identification in *Bougainvillea* spp. using SSR markers was carried out by Kumar *et al.*, (2014).They studied 25 cultivars namely 'Blondie', 'Chandrabieri', 'Cherry Blossom', 'Chitra', 'Dr. Bhabha', 'Dr. H.B. Singh', 'Shubhra', 'Singapore Red', 'Sweet Heart', 'Thimma', 'Lady Mary Baring', 'Mahara', 'Mahatma Gandhi', 'Mary Palmer Special', 'Mrs. Butt', 'Refulgens', 'Roosevelt's Delight', 'R.S.Bhatt', 'Sonnet', 'Spring Festival', 'Stanza', 'Summer Time', 'Vishaka', 'Zakiriana'. The PIC value varied widely among primers and ranged from 0.364-0.891 with an average of 0.761 per locus and the size of the amplified products ranged from 90bp -250bp. Primer BOUG-1 showed the highest polymorphism index content (0.891).Thus reflected its ability to differentiate these cultivars much better at molecular level. An unweighted pair group method cluster analysis (UPGMA) based on similarity values revealed five main clusters. The cluster-1 was the largest one comprising 18 cultivars while cluster IV and V emerged as the smallest ones comprising 3 cultivars each. This experiment was first of its kind in using microsatellite markers for phylogenetic analysis and molecular characterization of Bougainvillea cultivars. The study proved the efficiency of SSR markers in documentation, identification and tracing out molecular origin among unknown cultivars of Bougainvillea.

Rastogi *et.al.* (2015) studied on the 48 varieties of *Bougainvillea* for the determining genetic variability of these varieties. For genetic diversity study, they used two DNA fingerprinting methods *viz.* DAMD and ISSR. They found 91% of polymorphism amongst the *Bougainvillea* varieties. This polymorphism *i.e.* polymorphic information content (PIC) value was varied from 0.07-0.27 in DAMD and 0.13-0.34 in ISSR primers respectively. *Bougainvillea* varieties have a vast genetic variation. DAMD and ISSR analysis have identified the diverse *Bougainvillea* varieties which could be further utilized in various genetic improvement programmes including conventional as well as marker assisted breeding towards development of new and desirable *Bougainvillea* varieties. A comprehensive UPGMA dendrogram was constructed on the basis of combined data of DAMD and ISSR, which showed five major (A, B, C, D and E) clusters. Cluster A was the largest cluster among all the five main clusters and consisted 12 varieties. It was further divided into two sub-clusters AI (12 varieties) and AII (2 varieties). Cluster B was also divided into two sub-clusters viz. sub-clusters BI (9 varieties) and BII (3 varieties). Cluster C was the smallest cluster among the all five main clusters and comprised of only 3 varieties. Cluster D had two sub-clusters DI (6 varieties) and DII (3 varieties). Cluster E was further divided into two sub-clusters EI (3 varieties) and EII (2 varieties). It was also observed that 4 varieties did not show affinity to any of the five major clusters and their sub-clusters. The UPGMA (Unweighted Pair Group Method with Arithmetic Mean) dendrogram helped to know about the origin and parentage of these varieties. The lowest genetic distance was 0.17 and highest genetic distance was 0.60. The study was a comprehensive analysis of 48 varieties of bougainvillea and their genetic variability.

Therefore, molecular characterization was very successful for studying genetic diversity. This also helped identification and documentation of varieties and to trace out the molecular affinity of origin of unknown group of *Bougainvillea* varieties. Origin and interrelationship of different species / varieties of the four major species were studied successfully. Origin and affinity of varieties of unknown origin could be traced out up to certain level. Such study is very helpful and necessary for assessing genetic diversity of large germplasm collections of horticultural species and for their further improvement through selective breeding program.

*****

6

Culture and Management for Growth and Flowering

Bougainvilleas can be grown in a wide range of agro-climatic conditions. They are very hardy plant in comparison to other flowering perennials. However, vegetative growth and flowering varies according to the soil and climatic conditions. Therefore, it is essential to know the ideal agro-climatic conditions in order to get best growth and flowering.

Climate

Best performance of Bougainvilleas is achieved in sunny and hot climate. Though they can withstand wide range of climatic conditions, growth and development of bracts are affected under adverse conditions. They have been found growing well in sub-Himalayan tract to coastal areas. Thus, one can easily select them as flowering plant in the parks and gardens for all zones.

In north Indian conditions, best flowering period is during April–June. During this period, the average maximum day temperature ranges between 35-40^{0}C and the relative humidity 35-55%. These data indicate that high temperature together with low humidity is conducive for proper flowering of Bougainvilleas. Often it has been observed that climatic and physiological adversity by way of high temperature and moisture stress etc. induce better flowering.

Countries having tropical and sub-tropical climate is ideal for growth and flowering of bougainvilleas. For cold region, varieties of *B. glabra* are more suitable than *B. spectabilis*.

Soil

They can be grown in all types of soil. However, the soil should be well drained and rich in plant nutrients. Bougainvilleas do not grow well in wet soil as the water requirement is comparatively less. Slightly acidic soil having pH 6.0-6.5 is good for growth and flowering as it helps in availability of micro-nutrients. In all states of India, Bougainvilleas have been found growing well which indicates their wide adaptability. In most degraded soil where other flowering ornamentals failed to perform, Bougainvilleas grow and produce flowers well.

Planting Site

A sunny location away from the shade of buildings and trees is best for Bougainvilleas. If planted in locations having shade, less flowering and more vegetative growth has been observed. Development of colour of the bracts is also affected when planted in shade. Therefore, it is recommended to select sunny location for planting of Bougainvilleas.

Planting

Well grown plants having good vegetative growth are generally selected for planting. Every climatic regions of India have a particular planting time depending upon the season. In eastern and northern India, planting can be done during February to April and subsequently in July to September. In southern India, planting can be done almost round the year particularly when the temperature is not very high and level of humidity is adequate.

In other countries south-east Asia, planting may be done in moderate period keeping in mind favourable growth factors. Planting of Bougainvilleas can be done in various ways depending upon the purpose. The planting pattern, distance between plants and overall planning is usually done considering the objective of planting.

As Ground Plantation

Plantation in the ground is a very popular way of growing Bougainvilleas, as they perform very well when planted in the ground. The purpose of plantation is also varies depending upon the requirement. In ornamental gardens, ground plantation is done as specimen plant, along the fence / boundary wall, training over the gate, porch, pergola and several other ways. Alternatively, planting in groups is also done considering availability of space. In addition to that, plantation is done in formal or informal pattern in the form of a garden following a design and colour scheme. The main purpose is to display mass effect of bract colour and transformation of any place into a vivid landscape.

Pits and Planting - Pits measuring 60x60x60 cm are prepared at the specific points as per plan and purpose. The excavated soil is taken out and left for sun drying for 3-4 days. Subsequently, the soil is mixed with FYM in 3:1 ratio along with neem cake @ 250 g / pit. Insecticides (Thimate 10G / Furadon 3G) @ 10g / pit is also mixed with the soil before plating for controlling soil born insects. The mixed up soil is then filled in the pit and irrigated for consolidation.

Planting distances depend on the purpose and usually kept 2-3 m in between plants. Row to row distance is kept as per requirement of the planting scheme but not less than 2-3 m. Best period of planting is during monsoon when plants get easily established. However, places / countries having moderate climate, plantation may be taken up round the year or as per climatic condition. Extreme hot or cold seasons may be avoided for plantation.

Manures and Fertilizers – Well grown plants in the ground should be manured twice in a year, once in July and in February for encouraging growth and flowering. However, one should keep in mind the active growth period of the plants. In no case, manuring should not be done in extreme season. Bougainvilleas have fewer requirements of manures and fertilizers in comparison to other flowering perennial plants. Grown-up and well established plants do not require much application of manures and fertilizers. However, for best results an annual application of mixture of FYM (5-7 kg), bonemeal (250 g), neem cake (250g), superphosphate (250g) and potash (50 g) per plant is recommended. The mixture should be applied in the ground around at the basal ring followed by irrigation. Top dressing of the above mixtures may be done at regular intervals once in three months followed by forking of the plants.

Schedule of Irrigation - At the initial stage, young plants require irrigation for better establishment. Established plants in ground, however, require less watering. Irrigation during pre-flowering and flowering stage needs to be restricted which encourage profuse flowering. Fully grown and established plants in ground usually thrive well on rain water, unless otherwise required. However, the schedule of irrigation and frequency should be decided depending upon the season and soil moisture condition. In case of large number of plants in the ground, alternative irrigation methods viz. drip, Quick Release Coupling (QRC) based irrigation system should be installed by replacing manual irrigation system.

Pinching, Pruning and Training - The requirement and frequency of pinching, pruning and training depends on the use of the plant and its purpose. A different schedule is to be adopted for developing hedge, cascade, espalier, pot plant etc. If it is grown as climber on porch, arch, on stump, pruning and training becomes more essential and a repetitive process. Overgrown, dead and de-shaped branches are required to be pruned preferably before monsoon or after the flowering. However, intensity and frequency of the pruning is dependent on the varietal behavior and purpose of growing. Pruning and training complement each other for shaping the plants for various purposes.

Pot Culture

Bougainvilleas are successfully grown in pots or in large containers and produce flowers well. Over the years, these pots serve as a beautiful display item due to the colourful bracts. The main advantage pot culture is that the pots / containers can be placed for decoration purpose in any part of the gardens. Therefore, growing Bougainvilleas in pots / containers is another alternative way to use them in ornamental gardening.

Particularly, in small residential gardens where space is limited, pot grown ornamentals open up a new scope for beautification. Roof top gardens, balcony, portico where there is no ground space, potted Bougainvilleas are the first choice.

Pots / Containers - Type of pots / containers used for growing Bougainvilleas vary a lot. Selection depends on the purpose of use besides its placement in a particular location. The minimum size of the earthen pots/ cement pots should be 25-30 cm in diameter. Larger sized containers viz. cement pots, metal drums / wooden tubs / barrels having diameter 50-90 cm are also used for growing specimen plants for display in large gardens.

Potting Mixture - The composition of the potting mixture plays an important role in growth and flowering. Therefore, the mixture should be prepared properly by mixing garden soil and FYM in 3:1 ratio together with neem cake and bone meal @100 g / pot respectively.

Planting in pots / containers is done in moderate season of the year either in February-March or July-September, as the case may be according to the agro-climatic condition.

Recommended Varieties – Following varieties are recommended for growing in pots and containers – 'Aruna', 'Blondie', 'Cherry Blossom', 'Chitra', 'Dr. R.R. Pal', 'Isabel Greensmith', 'Lady Mary Baring', 'Mahara', 'Mary Palmer', 'Palekar', 'Roosevelt's Delight', 'Shubhra', 'Thimma' etc.

Pinching - Pinching of apical growth of the pot grown plants is also recommended for encouraging lateral branches and mass flowering. In addition, whenever required, pruning of the hardwoods shoud be done to keep the plants in shape.

Dwarfing - Dwarf and pot-bound Bougainvilleas are always an attraction for the garden lovers. Induction of dwarfing by using chemicals was remained a practice to fulfill the demand of dwarf plants. As per reports from experimental findings, certain chemicals are used as growth retardant viz. cycocel, maleic hydrazide, B-9, phospon etc. Sprays of maleic hydrazide @ 1000 ppm at weekly interval produced dwarf plants, as reported. Similarly, soil application of dust of the referred chemicals is also equally useful.

Manures and Fertilizers - Application and forking of manure mixture in pot / container grown plants, is recommended especially after pruning and in active growth period.

Excess application of manures and fertilizers leads to more vegetative growth and less flowering. Therefore, balanced application is recommended for overall best performance.

Irrigation - Plants grown in pots require regular watering, particularly at young stage, till they are established. Frequency of irrigation is decided according to season. Bougainvilleas are sensitive to overwatering. Therefore, irrigation schedule should be regulated as per requirement and considering the stage of the plant growth. During flowering season, restricted irrigation is done. Over irrigation, may result in to poor or lack of flowering.

Irrigation is usually done manually by watering can or hose pipe as per convenience. If the potted collection is in large number and the positions are fixed, drip irrigation method may be adopted.

Pruning Manual

Pruning refers removal of vegetative part (main branch, sub-branch and thin shoots) of the mother plant. The main purposes of pruning are as follows.

- Removal of unwanted branches/ twigs / shoots.
- To give the plant a proper shape depending upon the purpose.
- To encourage vegetative growth and flowering.
- Removal of broken and diseased branches.

Bougainvillea need pruning irrespective of climate and region for better flowering and to tame the unruly branches. It has been observed that well pruned plants perform in a better way with respect to vegetative growth and flowering than un-pruned plants. Formation of bracts of Bougainvillea usually takes place on current year new branches. Therefore, it is recommended that pruning should be done in order to obtain better overall performance.

Frequency and Intensity of Pruning - Intensity of pruning is an important factor as it regulates flowering which ultimately influences overall impact of the plant on the landscape. Removal of thick primary braches is required when the plant is hugely overgrown. The plants which have been pruned regularly in every year hardly require any removal of primary branches by way of deep pruning. Light pruning and pinching of the secondary branches are done regularly as per requirement.

Flowering Behaviour

Most of the Bougainvillea varieties produce flowers in succession round the year with varying degree of intensity. However, in north Indian plains peak season is March-April when nearly all varieties remain in bloom. It indicates that climatic conditions (minimum / maximum temperature 15-20° / 30-35°C; light intensity 80,000-1,00,000 Lux; Relative humidity 35-45%) during this period are conducive for flowering. Whole north Indian plains is subjected to prolonged cold temperature (maximum being 12-15°C) during winter season followed by fast rising of temperature in a March-May together with low relative humidity. It induces rapid growth and spurt of flowering. It has also been noted that bracts generally borne on new growth of current season.

The Problem - After peak flowering season is over, most of the varieties behave in different ways resulting scanty and very low flowering leading to un-presentable forms. Considering the fact that bracts generally appear on new growth, regular pruning is the only way to encourage new growth and flowering. With this objective, varieties need to be subjected to special schedule of pruning for induction of flowering.

Induction of Flowering by Clipping / Pruning – For increasing frequency and duration of flowering which is an important parameter for ornamental use, pruning is used as a tool for induction of flowering. In a study, 12 varieties ('Aida Variegata', 'Chitra', 'Lady Mary Baring', 'Mary Palmer', Mrs. Butt', 'Palekar', 'Poultoni', 'Poultoni Special', 'Shubhra', 'Tetra Mrs. McClean', 'Thimma', 'Zulu Queen') were selected which were growing in the ground. The observations were recorded over a period of one year during 2008-09. Each plant was provided with equal cultural treatments.

Single heavy pruning was done during July by removing all secondary green branches and keeping only matured hardwood. Subsequently, only green and thin branches were removed where bracts were borne. Thus, altogether nine such pruning was made after completion of each flush at an interval of 40 days. Observations were recorded on number of leaves per unit length of stem, number of bracts and their size in every month in order to assess flowering behaviour, frequency, intensity and finally, to identify good flowering varieties.

The technique of repeated light pruning after completion of each flush induced new growth and flowering in succession and with greater magnitude. Out of the 12 varieties 'Poultoni Special' was best considering frequency, duration and no. flowers borne followed by 'Palekar', 'Zulu Queen', 'Shubhra', 'Thimma', 'Lady Mary Baring' and 'Tetra Mrs. McClean'. These varieties along with the new pruning technique are recommended for extensive use in ornamental gardening.

Season - One should consider the season ahead, before taking up pruning operation. The intensity of pruning depends on the purpose as well as season. If main primary branches need to be pruned, monsoon or season having high humidity is best. Light pruning, pinching and removal apical meristem can be done in any season depending upon the requirement. In agro-climatic regions having moderate climate pruning may be done irrespective of the season.

Other Factors

The factors influencing flowering in Bougainvillea was studied by Ramina *et al..* (1979) at department of Environmental Horticulture, University of California. They observed that reproductive development, whether expressed as first node to flower or number of inflorescence developing is promoted in direct relationship to leaf area and in inverse relationship to the numbers of axillary branches developing. Soluble solids (%) in the reproductive shoots vary with the reproductive development. Development of inflorescence was promoted by cytokinin treatments and soluble solid (%) supported a nutritional regime in the control of flowering of Bougainvillea 'San Diego Red'.

Similar to the above work, control of flowering in Bougainvillea 'San Diego Red' as influenced by metabolism of benzylademine and the action of Gibberlic acid (GA) in relation to short day induction was studied by Even-Chen *et al..* (1979). They found that Benzyladmine (BA) and short day (SD) induction promoted and Gibberlic Acid (GA) inhibited flowering in Bougainvillea 'San Diego Red'.

Training Methods

There are several training methods for Bougainvilleas. The popular methods are – arch, pergola, bush, climber, cascade, espalier, hanging basket etc. Selection of suitable varieties as per training method is required for best performance.

Royal Daupline trained on wall   **Mass effect of colourful bracts**

Diseases and Pests

Bougainvilleas are very hardy and have strong defence mechanism. As a result, they are not attacked by diseases and pest in large scale. All over the world, there is no report regarding major diseases and pest which have been found to limit growth and flowering of Bougainvilleas. However, there are some report of minor diseases and pests which have been found affecting the plant occasionally.

Following is the brief account of diseases and pest of Bougainvilleas.

Fungal Diseases

A few fungal diseases have been reported from tropical parts of the world especially during high humid season.

Leaf Blight – This is a fungal disease caused by *Phytophthora parasitica*. This disease is characterized by development of lesions in small, irregular shape having ashy-green colour, usually starting from leaf tips and margins. It has been found that young leaves got infected more than the older leaves. Sometimes floral parts are also get infected. In advance stage of infection, the lesions increased in size and leaves became blackened and curled.

Spraying of copper fungicides (2%) is recommended as a control measure. Varieties like 'Glabra', 'Golden Glow', 'Thimma', 'Shubhra' etc. have been found resistant to leaf blight disease.

Leaf Spot – This also a fungal disease caused by *Cercosporitium bougainvilleum*. Dark brown to black stroma, globose to sub-globose in shape, pale brown in colour with moderately prominent conidial scars, straight or geniculate are the main diagnostic symptoms.

Spraying of copper fungicides (2%) is recommended as a control measure. *Bougainvillea* varieties under Glabra group have been found non-susceptible while varieties under Spectabilis group are moderately susceptible.

Bacterial Disease

Leaf spot – This disease was characterized by circular to irregular brown necrotic areas having yellow shade. The causal bacterium was identified as *Pseudomonas andropogonis*. Another bacterial leaf blight which developed symptoms as velvety pink spots has also been reported. The causal organism identified as *Septobacidium sp, Phyllosticta Bougainvilleae* and *Colletotrichum capsici*. However, the occurrence of bacterial leaf spot in *Bougainvillea* is not very common.

Viral Diseases

Bougainvillea spectabilis varieties have been found showing virus like symptoms by curling of leaf tip together with development of spiral yellow spots which was identified as viral disease by the presence of Bacilliform particles in leaf-dip preparation and by open reading frames sequence. The virus was tentatively named as '*Bougainvillea* Bacilliform Virus' (BBV).

In another report by Tsai *et. al.* (2005) from Department of Plant Pathology and Microbiology, National Taiwan University, Taiwan Bougainvillea Chlorotic Vein-banding Virus (BsCVBV), infecting Bougainvillea (*Bougainvillea spectabilis*) plants was observed as first report. Similar incidence of visrus diseases was also reported was first reported in Brazil in 2001. Infected bougainvillea plants developed symptoms such as mottling, chlorosis, vein-banding and stunting. Severe leaf-distortion symptoms were observed in the susceptible hybrid 'Taipei Red', the most popular bougainvillea cultivar in Taiwan.

Pests

Being extremely hardy, no major pests have been reported which attack Bougainvilleas in alarming scale and hamper vegetative growth and flowering. Sometimes, caterpillars, aphids, leaf miners, scale insects, thrips, white fly etc. have been found to attack the plant. The incidence of pest attack depends on the season, health of the particular plant as well as surrounding plot / field. These can be controlled by spraying of systemic insecticides as and when required. The spraying should be done in morning hours and in a sunny day.

*****

7

Different Methods of Multiplication

Generally, Bougainvilleas are multiplied by different vegetative methods for large scale production of plants for garden use as well as nursery trade. In specific cases, seeds are also used for raising plants. However, seed setting and production of seeds are dependent on the agro-climatic conditions as well as species.

Stem cuttings and layering are usually done for large scale propagation. Tissue culture is also practiced in limited scale especially for difficult to root varieties. The propagation method to be followed is dependent on the variety and its response to the particular method of propagation. The different multiplication methods usually practiced for Bougainvilleas are described below.

Stem Cuttings

This is most popular and commonly followed method of multiplication. Large member of varieties are propagated by stem cuttings and most successful method.

Hardwood Cuttings - Stem cuttings with hardwood having diameter of 0.75 to 1.5 cm are selected. A slanting cut is given at the tip of the cuttings while the basal portion is given round cut just below the node or inversely. Rooting of the cuttings is done in various ways either putting them in well prepared soil/sand beds or in containers under mist or otherwise. Stem cuttings are treated with auxin (hormone) powder like IAA (Indole Acetic Acid) or IBA (Indole-3-Butyric Acid) or NAA (1-Naphthalene Acetic Acid) either as commercial formulation or prepared solution (1000-4000 ppm). The varieties which are easily rooted with the help of auxin treatment have been reported as a result of experimental findings. The varieties are – 'Cherry Blossom', 'Dr. Rao', 'Dr. R.R. Pal', 'Glabra', 'Golden Glow', 'Lady Mary Baring', 'Los Banos Variegata', 'Mahara', 'Shubhra', 'Thimma', 'Zakiriana' etc.

Stem cuttings should be planted in mist house for better and successful rooting in large scale. Otherwise, rooting in open condition (in container/ beds) requires special attention regarding moisture management. Rooting usually takes place within 4-6 weeks. The rooting pattern of Bougainvillea varieties vary a lot. Majority of the varieties easily or moderately root and new plants are produced. But not all varieties behave in the similar way. Therefore, selection of varieties shall have to be made before going for hardwood stem cutting.

Cuttings with Softwood and Leaves - In this method, tip of the branches having softwood and a few leaves is selected as propagating material. Some varieties very successfully produce adventitious roots and new plants are raised as propagules. The best performance of softwood leaf cuttings can be achieved in mist-condition. The varieties which respond well in this method are – 'Cypheri', 'Dr. P.V. Sane', 'Mahara', 'Pradhan's Profusion', 'President Roosevelt', 'Refulgens', 'Splendens', 'Thimma' etc.

Air-Layering - This is another method of vegetative propagation suitable for those varieties which does not respond well to stem cuttings.

In Bougainvillea, air-layering is also practiced. Stems having pencil-thickness are selected for air-layering. A simple slanting cut is given over the bark up to the wood followed by wrapping with sphagnum moss, covered with polythene sheet and tied by cord. Root formation starts with in 2-4 weeks depending upon the variety. When adventitious roots are seen through polythene sheet, the stem is cut from the mother plant and planted in pots / containers as new plant. In northern and eastern India, the best season is monsoon (July-September) when air-layering is done. In southern India, due to high rainfall and prevailing moderate season, air layering may be done during March-October depending upon the season and variety.

In other tropical countries, the season having high humidity is ideal for air-layering.

Budding - This is an alternative method of propagation but not done in large scale. Varieties which do not respond well by stem cutting are usually propagated by budding. There are several varieties which are used as rootstock namely 'Dr. R.R. Pal', 'Glabra', 'Dr. H.B. Singh', 'Mrs. Butt' etc.

Shield budding and 'T' budding are commonly practiced. In north-India the best period for budding is March-April but can be done in any season except monsoon. In other countries, season having moderate temperature and humidity is best suitable.

Grafting

This is another method for multiplication for regeneration of new plants with specific objectives. Grafting is generally not practiced for the purpose of mass multiplication. The main purpose of grafting is development of novelty in multiple colours. These grafted plants are considered as prized items in horticultural nursery trade and fetch premium price.

Procedure – Grafting is done between suitable rootstock and selected scion of desirable varieties. In this process, mother plant of rootstock is selected first whether grown in pots or in ground. Not all Bougainvillea varieties respond well in the grafting. Therefore, selection of rootstock varieties is to be done having suitable branch of appropriate thickness. If required, excess branches are removed keeping only the selected branches for grafting. Afterwards, varieties which are to be grafted needs to be identified depending upon the bract size, colour and type. The rootstock should have sound health and strong root system.

Stems having 1.0-2.0 cm thickness containing axillary buds (no leaves) are selected for grafting purpose. The length of the stem piece may vary from 25.0-35.0 cm and should be straight. The basal portion of the stem is given slanting cut from two sides. The stem piece is inserted on the selected branch of the rootstock after making an insertion about 1.0-1.5 cm deep and tied with thin polythin strip so that entire grafted portion is covered.

Precautions and Aftercare – The mother rootstock should be maintained by proper intercultural operations for maintaining proper vegetative growth. Excess foliage and branches of the rootstock is removed so that growth initiation through grafted stem portion starts. Emergence of new shoots of the rootstock is repeatedly eliminated in order to direct the vegetative growth through grafted stem. Precaution should be taken to protect the plants from excess rains, heat and sun. Multiple grafting can be done in a single rootstock plant so that when in bloom the grafted plant show blooms in multiple colours.

Initiation of new growth from the axillary buds by way of emergence of new leaves indicates successful union of the grafting. Gradually, the grafted stem develops into a twig by the development of new branchlets and leaves. Finally, it starts blooming when the vegetative growth is completed. The grafted specimens containing multiple coloured blooms become a showpiece in the garden and enhance beauty.

Season for Propagation

India has got distinct climatic zones having specific temperature and rainfall pattern. Considering this, the propagation season for *Bougainvillea* differs. In the plains of northern India, vegetative propagation through stem cuttings is generally done during February-March. In southern India moderate season prevails round the year. The propagation through stem cutting may be done almost round the year.

Budding and grafting is generally done during moderate season in respective climatic zones when there is no rainfall. Layering is generally done in monsoon when high humidity prevails.

Seeds

Multiplication through seeds and raising plants is another method. However, this is not commonly practiced for large scale multiplication as seed production is limited and dependent on climate. As the plants raised from seeds may have different phenotypic characters, these may not be similar to the parents. Therefore, these are not very purposeful for production of true-to-type and for garden use.

However, for breeding purpose, plants raised from seeds have enormous importance. In fact, for creating new hybrids, seeds are important source. In the plains of northern India, seed setting usually takes place in April-May and seeds are sown in June-July germination of seeds start within 10-15 days. Varieties which bear seeds are - 'Lilac Puff', 'Lord Willingdon' 'Margery Lloyd', 'Red Triangle', 'Speciosa', 'Tetra Mrs. McClean', etc. Profuse seed setting have been observed southern India.

In other countries, seed sowing may be done in moderate season.

Tissue Culture

Tissue culture is a well-established technique for mass multiplication and production of true to type plantlets. In Bougainvillea also, tissue culture technique has successfully been used for multiplication purpose. It is very useful and successfully used for multiplication of difficult to root varieties.

As the demand of Bougainvillea is increasing day by day, the requirement of nursery plant is also going up. Propagation through stem cuttings alone can not meet the demand. In recent years, propagation of ornamental plants by tissue culture has become an essential commercial practice. Therefore, tissue culture is an effective alternative method of propagation of Bougainvillea. Many research scientists have worked on tissue culture of Bougainvillea for their standardization of protocol. It has been found that the response of different varieties under various species is variable and needs to standardize as per variety. Some experimental results on tissue culture of Bougainvillea have been cited here for use in commercial scale.

Work on shoot apex culture of *Bougainvillea glabra* 'Magnifica' was undertaken by Sharma *et. al.* (1980) in India. Shoot apices were introduced to regenerate an average of ten shoots from the base by using BAP (0.5 mg/l) with IAA (1.5 mg/l). Isolated shoots from such culture were rooted in a medium containing 0.1 mg/l each of IBA and 2,4,5-T and lacking BAP. Plantlets were grown successfully in pots and flowered normally.

In an experiment, Duhoky and Al-Mizory from Iran studied in vitro micropropagation of selected *Bougainvillea* species through callus induction. They developed a protocol for rapid callus induction and subsequent root generation in Bougainvillea (*B. buttiana, B. spectabilis* and *B. glabra*). The best result of callus induction response (92.86 %) in whole area of node explant was observed on WPM medium of *B. x buttina* with compared to *B. x buttiana* cultured in MS medium (85.71%). They got maximum number of shoots induced from callus, maximum shoot length and highest number of leaves per culture when WPM was used. The results were 12.14 shoots / culture, 2.14 cm shoot length and 20.71 leaves / culture from *B x buttiana* after 6 weeks of culture inoculation. In MS medium number of shoots induced from callus, maximum shoot length, and highest number of leaves per culture were 11.43 shoots, 17.14 cm, 1.96 numbers of leaves / culture respectively.

In vitro response of various growth regulators on the regeneration of *Bougainvillea spectabilis* was studied by Ahmad *et al.*. (2007) in Pakistan. They investigated optimum levels of BAP, g+ Glutaminelutamine, IAA and IBA which were supplemented to MS medium for microprogation. Maximum number of plantlets (81.25%) with conspicuous callus formation was noticed with BAP 1.0 ml/l + Glutamine 500mg/l followed by BAP1.0 mg/l+ 250mg/l Glutamine and 64.58 % plantlets were developed. Half strength of MS medium supplemented with IAA 0.5 mg/l and IBA 0.5mg/l was best medium with 79.16 % rooting.

In another experiment, micropropagation of *Bougainvillea spectabilis* 'Splendens' was taken up to standardize protocol to produce true-to-type plantlets. It was found that MS medium with BA 6.65 mg was suitable composition for axillary bud sprouting and shoot formation. Swamy and Sahijram (1988) worked on tissue culture propagation of Bougainvillea Indian Institute of Horticultural Research, Bangalore, India. They considered the problem of rooting of varieties like 'Mary Palmer', Louise Wathen', 'Cypheri', 'Laterietia' etc. which were difficult to root. Attempt was made to develop a micropropagation strategy of *B. glabra*. The experimental results showed that shoot tip grown on BMS supplemented with BA [8mg / l[1]+ NAA (4 mgl[1])] produced maximum number of multiple shoots. They also found that BA alone (4 mgl[1]) was also effective in the development of multiple shoots. The plantlets survived (77%) transplantation to soil in pot and population was raised successfully.

Steffen *et al..* (1988) experimented on growth and development of reproductive and vegetative tissues of Bougainvillea cultured in vitro as a function of carbohydrate. They worked on Bougainvillea 'San Diego Red' and cultured in media containing either 3% fructose, glucose or sucrose as carbon source. Growth and development of young leaves were equivalent whether sucrose or fructose was used. However, floret initiation on inflorescence meristem was much greater when fructose or glucose was the carbon source.

Mass propagation of *Bougainvillea spectabilis* through shoot tip culture was carried out by Shah *et al..* (2006) at Nuclear Institute for Food & Agriculture, Peshwar, Pakistan. They developed an in vitro regeneration protocol for *Bougainvillea spectabilis* by using shoot tips from 5 year old plants. Culture of shoot tip was done in MS medium supplemented with different concentrations of BAP (0.25 – 2.0 mg/l) or Kinetin (0.25 – 2.0 mg /l) and NAA (0.1-0.5 mh/l) in combination with BAP (0.25 – 5.0 mg/l). It was reported by them that BAP (0.25 mg/l) combined with 0.1 mg/l NAA gave best result with 90 % shoot development and transformation into plantlet. The plantlets after weaning and acclimatization were transferred in pots and successfully developed into plants.

Generally Bougainvilleas are multiplied vegetatively by cuttings which has certain limitations so as to survival percentage as well as less response of certain varieties. Therefore, tissue culture is proven tool for quick and mass propagation of Bougainvilleas for commercial purpose.

Potted mother plants

Propagated plants in different stages

Revival of old plants by grafting

Sprouting of scion after grafting

Stem cuttings   **Varietywise display for sale purpose**

Establishment of Bougainvillea Nursery

Bougainvillea is a versatile plant of great ornamental value and very popular among the garden lovers. Its wide array of bract colour produces an amazing effect on the landscape which could not be achieved by any other flowering ornamentals. Be it a pot plant or standard or bonsai or trained over arch, all forms of Bougainvilleas are unique. Variation of bract colour together with forms (single / double) is the main reasons of attraction. Considering its huge demand and short supply, it is recommended that establishment of nurseries exclusively for commercial production of Bougainvilleas is to be popularized.

Availability of authentic germplasm collections are now localized and possessed by some R & D institutions, societies and some progressive nurseries where large scale multiplication is not possible due to the scientific nature as well as other constraints. Therefore, efforts must be initiated to encourage either new people of the semi-urban areas or some of the established nurserymen to take up this profitable entity.

Commercial Potentiality

Bougainvilleas are high in demand due to large scale plantation in urban areas due to its draught tolerant capacity and low maintenance requirement in comparison to other plants. Potted plants for display, ground plantation in large public parks, plantation in road dividers, along the boundary wall of factories have created a sharp rise in the demand of Bougainvilleas. Therefore, there is enough potentiality for its further commercialization by way of setting new nurseries for mass multiplication using modern methods viz. mist house and use of growth hormones.

Nursery Establishment

Considering its commercial potentiality, setting Bougainvillea nursery will be of successful profitable unit. People who have land in semi-urban areas may take up this business in small scale initially and gradually expanding the same into a full-fledged nursery. Initially one acre of land may be earmarked for development of the nursery.

Layout of the Nursery – It is recommended that a layout plan is to be prepared for placing of different features in a proper way. This will help in proper placement of the features, better utilization of land and operational convenience. The features - road, path, propagation beds, store, mist house, tube well, potting shed, mother plant block, propagation block, display area etc. have to be listed out as per requirement and are to be marked in the layout plan (see plan).

Collection of Mother Plants – In horticultural nursery business genuine varieties with proper name is very important. For achieving this, it is recommended that mother plants are to be collected from genuine sources so that named varieties can be produced subsequently. During collection, all plants have to be lavelled properly and after planting record are to be maintained so that there is no mixing afterwards.

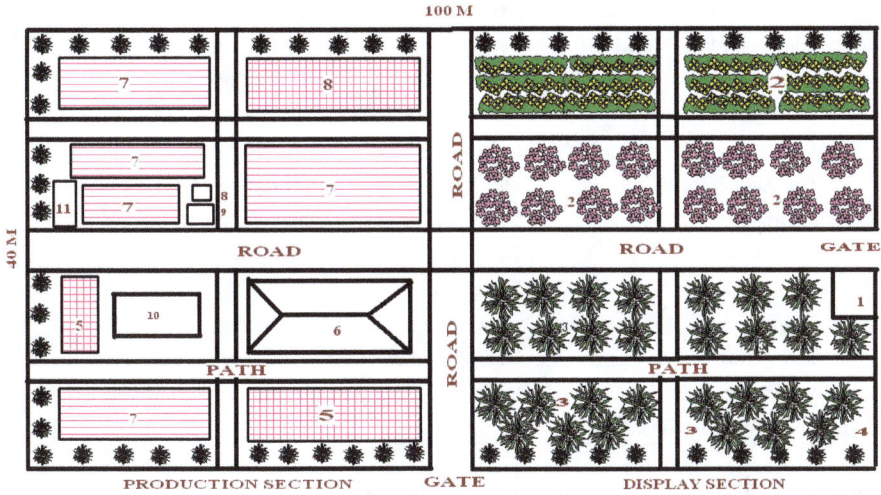

LAYOUT PLAN OF A MODEL BOUGAINVILLEA NURSERY

[Not to Scale]
[CAD by Dr. R.K.Ray]

LEGEND

1. OFFICE & SALE COUNTER	6. MIST HOUSE FOR PROPAGATION
2. PLANTS FOR SALE	7. PROPAGATION BEDS
3. ELITE MOTHER PLANT BLOCK	8. TUBE WELL
4. OTHER MOTHER PLANTS	9. STORE
5. PLNATS IN POLY BAGS	10. POTTING SHED
	11. MANURE PIT

Planting and Maintenance – Plantation of mother plants in proper way is of utmost importance. Pits measuring 60x60x60 cm are to be dug out at desired points as per layout plan. Dug out soil is left for a week for sun drying, later mixed with cow dung manure (one basket in each pit) and mixed with the soil together with Thimate 10 G @ 50 gm per pit. Best time of planting is monsoon when plants easily get established.

Propagation Methods – There are main two methods of propagation – stem cuttings and air layering. Bougainvillea cultivars are categorized into two groups depending upon their rooting habit *viz.* easy-to-root and difficult-to-root. Stem cuttings (matured) are suitable for easy-to-root types while softwood leaf cuttings and air layering are followed for difficult-to-root types. In moderate climatic regions, propagation can be taken up throughout the year but in other zones it should be done either in January-February or during July-August (monsoon). Under Mist House stem / leaf cuttings can be raised round the year.

S. NO.	Name of the Cultivar	Category	Recommended Propagation Method
1.	'Blondie', 'Chitra'	Easy-to- Root	Stem Cuttings (Hardwood)
2.	'Dream', 'Dr. Harbhajan Singh'	Easy-to- Root	Stem Cuttings
3.	'Dr.R.R. Pal', 'Flame'	Easy-to- Root	Stem Cuttings
4.	'Enid Lancaster' ' Flame'	Easy-to- Root	Stem Cuttings
5.	Mrs. Butt', 'Mrs. H.C. Buck'	Easy-to- Root	Stem Cuttings
6.	'Princess Margaret Rose'	Easy-to- Root	Stem Cuttings
7.	'Palekar' , 'Partha'	Easy-to- Root	Stem Cuttings
8.	'Poultoni', 'Shubhra'	Easy-to- Root	Stem Cuttings
9.	' Thimma', 'Mary Palmer'	Easy-to- Root	Stem Cuttings
10.	'Zulu Queen'	Easy-to- Root	Stem Cuttings
11.	'Splendens'	Difficult-to- Root	Air Layering
12.	'Red Triangle'	Difficult-to- Root	Air Layering
13.	'Pradhan's Profusion'	Difficult-to- Root	Air Layering
14.	'Vitthal'	Difficult-to- Root	Air Layering
15.	'Lateritia'	Difficult-to- Root	Air Layering
16.	'Lord Willingdon'	Difficult-to- Root	Air Layering
17.	'Margery Lloyd'	Difficult-to- Root	Air Layering
18.	'Bois de Rose'	Difficult-to- Root	Air Layering
19.	'Scarlet Queen Variegata'	Difficult-to- Root	Air Layering
20.	'Louise Wathen Variegata'		

Raising of Large Specimen Plants for Display – At present the demand for large specimen plants are growing day by day for producing immediate effect in the landscaping. In view of that specimen plant of bougainvillea in large cement pots/large P.V.C. pots/containers having 60-75 cm diameter are to be raised in the nursery. These plants fetch much more price than those raised in poly bags. Selection of good floriferous varieties like 'Shubhra', 'Mary Palmer Special', 'Thimma', 'Palekar', 'Poultoni Special' 'Dr. H.B. Singh', 'Zulu Queen', 'Land Mary Baring', 'Scarlet Queen Variegata' etc. are preferred for this purpose.

Raising of Plants for Planting – Another set of propagated plants is recommended for transplanting in poly bags of different size aiming for ready to sale, hold & sale for future. Make-shift / foldable mist irrigation facility with sprinkler should be used for economizing water and manpower.

Source of Plant Material in India – Initial collection of mother plants should be made from following reliable nurseries and reputed organizations.

    1. CSIR-National Botanical Research institute, Lucknow

    2. ICAR-Indian Agricultural Research Institute, New Delhi

    3. Bhabha Atomic Research Institute, Trombay, Mumbai

    4. ICAR- Indian Institute of Horticultural Research, Bangalore

    5. The Agri-Horticultural Society of India, 1,Alipore Road, Kolkata

    6. Madras Agri-Horticultural Society

7. KSG's Farm & Nursery, Alwarpet, Chennai

8. Lal Bagh Botanic Garden, Lal Bagh, Bangalore

9. The Nurserymen Cooperative Society Ltd., Lal Bagh Bangalore

10. Govt. Sunder Nursery, Nizamuddin, New Delhi

Techno-economics

Establishment of Bougainvillea Nursery and its sustainability as a commercial entity are dependent on following factors.

Selection of proper varieties – Such cultivars are to be selected which have demand and popularity in the market demand.

Optimization of land use – Perfect planning while laying out of the nursery so that best utilization of land is possible.

Cost-effective management of recurring and non-recurring expenditures – Initially fencing may be made by bamboo poles instead of iron angles for reducing cost. Mist house and potting shade may also be made low-cost type. Main aim should be minimizing recurring as well as non-recurring cost for maximizing profit.

Effective marketing chain and outlets – This is most important and principal link for commercial viability. Nursery owner must be in contact with the leading outlets in the urban areas for proper marketing both in retail and bulk.

Addition of new and novel cultivars year after year – New and novel cultivars are always remain an attraction for Bougainvilleas growers. Therefore, introduction of new cultivars in the nursery is a must.

Different heads of expenditures for establishment of the nursery have been listed out and categorized in the following table.

S. No.	Category	Items
1.0	Non-Recurring Items / Expenditures	Fencing, Road/Path
1.1		Tube well, Irrigation Net work
1.2		Electricity, lighting
1.3		Mist House, Potting Shed
1.4		Store, Office, Sale Counter
1.5		Initial stock of mother plants
2.0	Recurring Items / Expenditures	
2.1		Tools & Implements
2.2		Insecticides & Fungicides
2.3		Chemicals (Hormones)
2.4		Manures & Fertilizers
2.5		Mother Plants (For enrichment)
2.6		Manpower
2.7		Pots & Containers
2.8		Miscellaneous (Labels, markers, rope, threads etc)

Tentative estimates of the expenditures for production of different varieties and sale price along with profit percentage have been provided so that one can see the profitability in this venture.

S. No.	Name of the varieties of Bougainvillea	Category of the Varieties	Av. Market Price / plant (INR)	Av. Cost of Production / plant (INR)	Approx. Profit in INR (%)
1.	'Chitra', 'Thimma', 'Archana', 'Refulgens', 'Mahara'.	Elite	150.00	60.00	150
2.	'Scarlet Queen Variegata', 'Arjuna'	Elite	150.00	60.00	150
3.	'Los Banos Variegata', 'Dr.P.V.Sane'.	Elite	150.00	60.00	150
4.	'Splendens', 'Parthasarthy', 'Roseville's Delight', 'Pixie variegata'.	Elite	150.00	60.00	150
5.	'Tetra Mrs. Mc Clean', 'Mary Palmer Special', 'Cherry Blossom'.	Elite	150.00	60.00	150
6.	'Dr. Harbhajan Singh', 'Dream', 'Chitra'.	Standard	125.00	50.00	150
7.	'Lady Mary Baring', 'Glabra', 'Poultoni Special'.	Standard	125.00	50.00	150
8.	'Mahara', 'Mary Palmer Special', 'Parthasarthy'.	Standard	125.00	50.00	150
9.	'Roseville's Delight, 'Cherry Blossom', 'Poultoni Special'	Standard	125.00	50.00	150
10.	'Shubhra', 'Zulu Queen'	Standard	125.00	50.00	150
11.	'Mrs. Butt', 'Mrs. H.C. Buck'	Common	100.00	40.00	150
12.	'Palekar', 'Dr. R.R.Pal', 'Flame'	Common	100.00	40.00	150
13.	'Arjuna', 'Palekar', 'Glabra', 'Gopal'	Common	100.00	40.00	150
14.	'Dream', ' Tomato Red', 'Pixie'	Common	100.00	40.00	150
15	Jayalaxmi, 'Isabel Green Smith'	Common	100.00	40.00	150

*****

8

Cytology, Breeding and Development of New varieties

Cytology and Breeding

Cytological studies are important tool to know the structure, composition and other details of the cells. This also helps to select proper breeding methods for the varieties. In Bougainvillea, not much cytological work has been done worldwide. Important cytological work was done by Cooper (1931), Panch *et. al.* (1960) besides Khoshoo and Zadoo (1969). They suggested that most of the *Bougainvillea* varieties are diploids (2n=34) while a few are triploids (2n=51) and tetraploids (2n=68). Triploid varieties in existence are - 'Cypheri', 'Temple Fire', 'Lateritia', 'Poultoni Special', 'Perfection' etc. Naturally occurring haploid variety ('Pequinito') has also been reported. Artificially induced tetraploid varieties viz. 'Mrs. McClean', 'President Roosevelt' was developed by Khoshoo and Zadoo (1969). They also developed anuploids by crossing a triploid [2n=36 (2x +2) X 2n=40 (2x +6)].

Review

A cytological study comprising of 90 varieties was done by Zadoo *et. al.* (1975 d). They reported that all taxa were diploid (2n=34) except 'Perfection' and 'Poultoni Special' which were triploid. It was further stated that varieties of *B. glabra* and *B. spectabilis* shown 17 II and meiotic metaphase, while subsequent stages was normal leading to 80-90 % pollen stainability. Further, *B. glabra-peruviana* and *B. specto-peruviana* showed reduced chromosome pairing while *B. specto-glabra* showed normal pairing and disjunction. On the basis of the findings, parental species were differentiated at sub-chromosomal level as duplicate/ deficient gametes and resulted partial sterility in the hybrids even after normal pairing the triploid varieties developed as a result of formation of diploid gametes due to precious centromere division at metaphase (Ohri and Zadoo, 1986).

Holttum (1955) reported self-sterility of *Bougainvillea* varieties along with Khoshoo and Zadoo (1969). However, development of seeds and hybrids by selfing was reported from some parts of India especially from moderate climatic regions. It was found that seed setting was greatly influenced by the environmental factors besides health and age of the plants. In southern India, due to prevailing moderate climate, seed setting was frequent especially during the period February to April. However, in north India seed formation was poor due to severity of winter and summer.

Induced tetraploidy and restoration of fertility in sterile varieties of Bougainvillea was carried out by Zadoo *et. al.* (1975). Like fertile diploid varieties, induced tetraploids were self-incompatable but on crossing set seeds readily. Diploid progenitors showed irregular meiosis but predominent bivalent pairing was there in tetraploid counterparts. Lower number of chiasmata per chrosome in diploids in comparison with tetraploid counterparts was observed. Predominent bivalent pairing accompanied by high fertility was a strong pointer towards preferential pairing in induced tetraploids.

Pollination mechanism and breeding system under the studies of cytogenetics of cultivated Bougainvillea was done by Zadoo *et. al.* (1975). In Bougainvillea varieties, floral morphology together with coincidence of stigma receptivity and anther dehiscence suggests a self-pollinating mechanism. The study showed that about 80% of garden Bougainvilleas are sterile both in male and female. The sterile forms do not set seeds on selfing but do so readily on crossing. Therefore, for effecting breeding system, pollination mechanism together with anther dehiscence and stigma receptaibilty need to be studied particularly for varieties under cultivated species.

In another observation, Khoshoo and Zadoo (1969) reported high percentage of seed formation when tetraploid varieties were crossed with diploids, triploids and tetraploids.

Development of New Varieties

There are several methods for the development of new varieties namely hybridization, mutation breeding (physical and chemical). Spontaneous mutations in the form of bud sport are another way to get new varieties after selection and multiplication.

Bud Sport / Spontaneous Mutation

This is a natural method in which mutations are created by sunlight without any human interference. Sometimes, morphological changes of flower colour, foliage aberration or any other change is created by spontaneous mutations in the form of bud sport. It is required to spot these natural mutants out of the whole germplasm collections. Identified bud sports are isolated, multiplied vegetatively. Subsequently, performance is evaluated in the successive generations with regard to the stability of changed morphological characters and its distinctiveness in comparison to the parental characters. This is most simple method and remained a source of development of many new varieties in the past. Bougainvillea varieties developed by this method are – 'Abhimanyu', 'Arjuna', 'Aruna', 'Archana', 'Bhabha', 'Cherry Blossom', 'Jawahar Lal Nehru', 'Manohar Chandra Variegata', 'Parthasarthy', 'Scarlet Queen Variegata', 'Surekha', 'Thimma' etc.

Hybridization

This method is very successful for the development of new varieties of Bougainvillea by crossing male and female parents. Sometimes, there is natural hybridization also when large numbers of parents are kept together. New hybrids, thus got, need to be assessed and multiplied.

Planned cross breeding at the initial stage was very few and taken up in isolated way. In one such case, James Hendrey, Everglad Nursery, Florida made hybridization involving 'Rosa Catalina' (female) and 'Lateritia' (male). Two varieties namely 'Daniel Bacon' and 'Margaret Bacon' were developed. Similarly, another firm M/S. W.N. Stands developed two varieties 'Lady Seton James' and 'Lady Watts' as a result of hybridization between 'Sanderiana' x 'Lateritia' and 'Rosa Caralina' x 'Lateritia'.

Technique of Hybridization

The details of hybridization technique, specifically for Bougainvillea, involving emasculation and pollination was elaborated by German (1954), Percy-Lancaster (1959), Zadoo and Khoshoo (1975). Usually, the flowers appear in a group of three, each subtended by a prominent bract. Structurally flowers are tubular, slightly constricted in the middle with swollen base and funnel shaped upper part. The flower tubes usually have five ridges along the length ending to the terminal star. The carpel is single and encircled by ring shaped nectar. Stamens arise from the rims (Zadoo *et. al.*, 1975 b). Flower bud is split opened longitudinally before they open for breeding purpose and emasculated with the help of forceps / needle. All other parts are removed keeping intact the stigma without any injury. The best time for emasculation is morning. Flowers of the donor parent from which pollens will be collected were also bagged. Pollination is done in the next morning. After removal of the flowers, pollens are gently dusted over the stigma of the emasculated flowers. After pollination, the flowers are again bagged to prevent natural crossing. If successfully pollinated, the ovary enlarges and formation of seeds starts. The seeds are similar to a wheat grain. Seeds are sown immediately after maturation. Seed sowing to flowering usually takes about a year.

Notable *Bougainvillea* varieties developed by hybridization at CSIR-NBRI, Lucknow, India are – 'Begum Sikander',1969 [Dr.B.P.Pal x 'Jennifer Fernie'], 'Chitra', 1981 ['Tetra Mrs. McClean' x 'Dr.B.P.Pal'], 'Mary Palmer Special', 1974 ['Princess Margaret Rose' x 'Dr.B.P.Pal'] and 'Wajid Ali Shah', 1974 ['Dr.B.P.Pal x 'Mrs. Chico'].

Mutation Breeding

In order to create genetic variability within a crop, several plant breeding methods are employed. Hybridization is the main breeding method used by the plant breeders to combine desirable characters from various sources into one genotype. On the other hand, mutation breeding is an established and alternative method for development of new varieties, particularly in case of Bougainvillea. By this method plant genes are altered by treating seeds or any other plant parts with chemical or physical mutagens.

Physical Mutagenesis

In physical mutagenesis gamma radiation (cobalt 60) is used as mutagen. As Bougainvilleas are multiplied vegetatively, they are highly suitable for radiation by gamma rays. A large number of varieties have been developed by this technique.

The Technique - Usually stem cuttings in required length are irradiated with different doses. Usually the doses varies from 0.5 Kr (1 Kr = 10 GY) to 2.5 Kr which have been found to create mutants. Irradiation with higher doses has been found to produce non-useful abnormalities. After treatment, these cuttings are planted in beds or containers for rooting and finally growth performance is studied. Screening of the whole treated population is done carefully to identify desirable mutants. In case of development of desirable mutant, the same is isolated and further multiplied vegitatively to raise population having same genotypic and phenotypic characters.

Some varieties of Bougainvillea developed by gamma radiation in India are – 'Arjuna', 'Los Banos Variegata', 'Los Banos Variegata Silver Margin', 'Mahara Variegata', 'Abnormal Leaves', 'P.V. Sane' etc.

Chemical Mutagenesis

Several chemicals are used as chemical mutagen for inducing genetic variability as mutants. Chemicals which are commonly used for such treatments are - colchicines, ethyl methane sulphonate (EMS: 0.01 - 0.03%) and methyl methane sulphonate (MMS). Mainly aqueous solution of these chemicals are used in various concentrations and applied over the current vegetative growth (meristematic cells) for creating genetic variability either in vegetative part or in the flowers. The mutants thus developed are separated out subsequently, multiplied and ultimately developed into new variety after evaluation through standard protocol (Jayanti and Datta, 2006).

Varieties developed through chemical mutagenesis in India are –'Los Banos Variegata Jayanti' and 'Pixie Variegata'.

Significant Contribution of Research Institutes/Universities/Organizations of India for the Maintenance of Germpalsm Collection of Bougainvillea and Development of New Varieties

1. CSIR-NBRI (National Botanical Research Institute, Lucknow, India)

[A premium botanical research institute in India for multi-disciplinary botanical research work and germpalsm collection].

CSIR-NBRI initiated R&D work during 1960s considering ornamental importance and commercial potentiality of Bougainvilleas. Building up of germplasm collection was started, as a first step, in the botanic garden for creating a broad genetic base for taking up research and improvement work. In India, CSIR-NBRI was the first institute to take up research work on Bougainvilleas. Collection of varieties was initiated from various sources and conserved as living specimens in the form of bougainvillea garden and potted collection. At present, the institute has collection of four basal species of *Bougainvillea* (*B. spectabilis*, *B. glabra*, *B. peruviana* and *B. x buttiana*) and 200 named varieties. Moreover, these collections are often referred for identification, research and improvement work on Bougainvilleas taken up by National and International institutions/ organizations.

Germplasm Collection

The living germplasm collection of Bougainvilleas has been maintained in the following two ways.

Potted Collection – The collections have been displayed in an enclosure measuring 50 x 8 m. The varieties have been arranged and displayed in alphabetical order keeping three potted plants together. Altogether, there are 600 pots in the collection having plant labels so that one can easily identify and locate the varieties.

As Bougainvillea Garden - This is spread over in an area of one acre. There are 165 varieties planted in this garden as per specific layout plan. There are four sectors planted with four horticulturally important species viz. *B. glabra, B. spectabilis, B. peruviana* and *B. x buttiana.* Under these species different varieties have been planted in rows at a distance of 5 m. All the varieties are labelled for their educative and aesthetic use.

Development of New Varieties

The live germplasm collection, which was developed over the years, served as a broad genetic base. Moreover, the collection was used for multidisciplinary research work for the development of new varieties through hybridization, selection, bud sport, polyploidy and mutation breeding. As a result, a large number of new and novel varieties have been developed by the institute. Altogether, 26 new varieties have been developed by employing different plant breeding methods (Table 8.1)

Table 8.1. List of new varieties developed by CSIR-NBRI

S. No.	Name of the Variety	Methods	Parent	Year of Development
1	'Begum Sikander'	Hybridization	'Dr. B.P. Pal' x 'Jennifer Fernie'	1969
2	'Shubhra'	Bud Sport or Spontaneous Mutation	'Mary Palmer'	1969
3	'Dr. B.P. Pal'	Colchiploidy	'Shubhra'	1969
4	'Tetra Mrs. McClean'	Colchiploidy	'Mrs. McClean'	1969
5	'Arjuna'	Induced Mutation (Gamma Radiation)	'Partha'	1972
6	'Archana'	Bud Sport or Spontaneous Mutation	'Roseville's Delight'	1973
7	'Mary Palmer Special'	Hybridization	'Princess Margaret Rose' x 'Dr. B.P. Pal'	1974
8	'Wajid Ali Shah'	Hybridization	'Dr. B.P. Pal' x 'Mrs. Chico'	1974
9	'Parthasarthy'	Bud Sport or Spontaneous Mutation	'Partha'	1977
10	'Shweta'	Bud Sport or Spontaneous Mutation	'Trinidad'	1979
11	'Chitra'	Hybridization	'Tetra Mrs. McClean' x 'Dr. B.P. Pal'	1981
12	'Surekha'	Bud Sport or Spontaneous Mutation	'Scarlet Queen'	1981
13	'Nirmal'	Bud Sport or Spontaneous Mutation	'Mrs. McClean'	1982

14	'Manohar Chandra Variegata'	Bud Sport or Spontaneous Mutation	'Manohar Chandra'	1985
15	'Pallavi'	Induced Mutation (Gamma Radiation)	'Roseville's Delight'	1987
16	'Hawaiian Beauty'	Bud Sport or Spontaneous Mutation	'Hawaiin White'	1990
17	'Los Banos Variegata'	Induced Mutation (Gamma Radiation)	'Los Banos Beauty'	1990
18	'Mahara Variegata'	Induced Mutation (Gamma Radiation)	'Mahara'	1994
19	'Los Banos Variegata - Silver Margin'	Induced Mutation (Gamma Radiation)	'Los Banos Beauty'	2002
20	'Mahara Variegata Abnormal Leaves'	Induced Mutation (Gamma Radiation)	'Mahara'	2002
21	'Los Banos Variegata Jayanti'	Chemical Mutagen Induced Mutants	'Los Banos Beauty'	2006
22	'Aruna'	Induced Mutation (Gamma Radiation)	'Palekar'	2008
23	'Pixie Variegata'	Chemical Mutagen Induced Mutants	'Pixie'	2009
24	'Abhimanyu'	Bud Sport or Spontaneous Mutation	'Arjuna'	2010
25	'Dr. P.V.Sane'	Induced Mutation (Gamma Radiation)	'Dr. R.R. Pal'	2011
26	'Dr. A.P.J.Abdul Kalam'	Bud Sport	'Fantasi'	2015

NBRI- Abhimanyu

NBRI-Arjuna

NBRI-Aruna

NBRI-Begum Sikander

NBRI-Chitra

NBRI-Dr. P. V. Sane

NBRI-Hawaiian Beauty

NBRI-Jayanti

NBRI-Mary Palmer Special

NBRI-Nirmal

NBRI-Pallavi

NBRI-Parthasarthy

NBRI-Shubhra

NBRI-Shweta

NBRI-Tetra Mrs. Mc Clean

NBRI-Manohar Chandra Variegata

NBRI-Pixie Variegata

NBRI-Los Banos Variegata

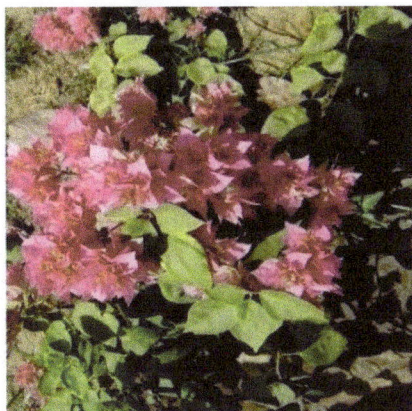

Silver Margin NBRI-Los Banos Variegata

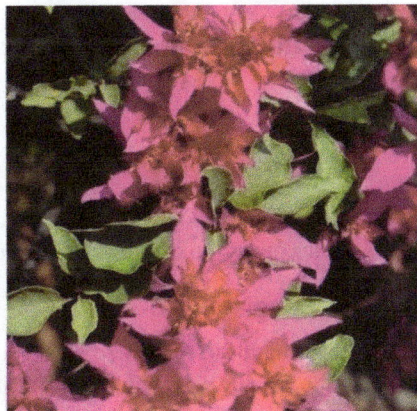

NBRI-Mahara Abnormal

Multiplication of Authentic Plant Material

Named varieties Bougainvillea has got a good demand especially in the nursery trade as well as for research work usually taken up by R&D institutions and universities. Keeping this in view, large scale multiplication of the available germplasm collection was taken up. As a result authentic planting material was produced which were provided to the research institutes and Bougainvillea lovers. Over the years, the germplasm collection of Bougainvilleas maintained in the botanic garden, CSIR-NBRI, Lucknow has been a source of well identified plant material thought the India and in cases abroad also.

Popularization

Popularization of Bougainvilleas as a flowering garden plant was taken up by this institute along with the R&D work since 1960s. Selected germplasm collections as well as newly developed varieties were displayed in the Flower Shows organized by the institute. Over and above, potted collection was exhibited in the State Flower Shows as well as Bougainvillea Festival organized by various Departments / organizations. The visitors in general and *Bougainvillea* growers in particular were made aware about the ornamental importance, pollution tolerance ability and low maintenance requirement of the Bougainvilleas besides their varied usage in ornamental gardening.

Registration of New Varieties

New varieties of Bougainvilleas developed by CSIR-NBRI have been registered with the International Registration Authority, IARI, New Delhi and Protection of Plant Varieties and Farmers' Right Authority (PPV&FRA), Govt. of India. The new varieties developed by any individual / nurserymen/ institutions / organization can be registered with the both authorities.

Recognition

The live germplasm collection maintained by the institue and the new varieties developed by CSIR-NBRI has been recognized as 'National Collection' by the National Biodiversity Board, Ministry of Environment, Forest, wildlife & Climate Change and Ministry of Agriculture, Govt. of India.

ICAR-IARI (Indian Council of Agricultural Research – Indian Agricultural Research Institute, New Delhi, India)

Role of IARI

IARI is a pioneering centre for germplasm collection, research and breeding of Bougainvilleas at the Division of Floriculture and Landscaping. Bougainvillea was originated in South America but approximately 50% of the present day cultivars were developed in India.

International Registration

The institute has been recognized as the International Centre for Registration Authority for Bougainvilleas (ICRA) by Royal Horticultural Society, U.K, since 1966. The division is also one of the major repository centres and wide range of genotypes of bougainvilleas is conserved. An international Check-list has been published by this Division which is an official document of variety developed and registered.

It is also a DUS (Distinct, Uniform and Stability) test centre of Bougainvillea recognized by Ministry of Agriculture & Farmers' Welfare, Govt. of India.

New Varieties

The new varieties developed by IARI are – 'Dr. R.R. Pal', 'Stanza', 'Summer Time', 'Sonnet', 'Spring Festival' and 'Vishakha'. A walk and learn path has been developed for easy identification of varieties and species at the repository for the benefit of the students and the visitors.

ICAR-IIHR (Indian Council of Agricultural Research – Indian Institute of Horticultural Research, Bangalore, India)

Role of IIHR

Maintaining live germplasm collection of Bougainvillea for research and ornamental purpose. Undertakes breeding work for the development of new varieties.

It is also a DUS (Distinct, Uniform and Stability) test centre of Bougainvillea recognized by Ministry of Agriculture & Farmers' Welfare, Govt. of India.

New Varieties

Following new varieties have been developed by IIHR – 'Chitravati', 'Dr. H.B. Singh', 'Jawaharlal Nehru', 'Purple Wonder', 'Sholay', 'Usha'.

First Patented Bougainvillea Variety

The first Indian variety of ornamental plant that secured a foreign patent is Bougainvillea. A variety, Bougainvillea 'Dr. H.B. Singh', developed at the IIHR, Bangalore won admiration in Australia. It was patented in Australia in the name of 'Krishna'. The variety is exclusively used as a potted plant and gained considerable popularity.

S/CAUs (State & Central Agricultural Universities of India)

In India, there are around 50 state and central agricultural universities located in various parts of the country. Under the subject horticultural research, Bougainvillea is tought and grown as an ornamental crop. Live collection is maintained besides research work on various aspects is also carried out including development of new varieties. By that way, the S/CAUs contribute significantly for the popularization of Bougainvillea.

CPWD (Central Public Works Department)

This Govt. Department has been doing a magnificent work by maintaining large parks in India where Bougainvilleas are grown as ornamental plants in various ways. Moreover, Bougainvilleas are planted in road dividers as well as mixed plantation along the avenues being hardy and pollution tolerant. Therefore, display of Bougainvillea varieties and their popularization in ornamental gardening has taken care of by CPWD.

BARC (Bhabha Atomic Research Centre, Mumbai, India)

The BARC, India is the premier nuclear research centre surrounded by the Trombay hills on three sides with the fourth side facing the Arabian Sea. In addition to the nuclear research, the centre has a unique collection of ornamental plants specially native trees, bougainvillea, canna etc.

Bougainvillea Garden – A germplasm collection of well indentified 150 varieties laid out in an informal way is centre of attraction of the Bougainvillea growers.

New Varieties Developed – The centre has developed four new varieties *viz.* 'Suverna', 'Silver Top' Jaya, 'Jayalakshmi Variegata'.

Bougainvillea Garden, BARC

IIHR-Spring Festival

IIHR-Stanza

IIHR-Summer Time

IIHR-Vishakha

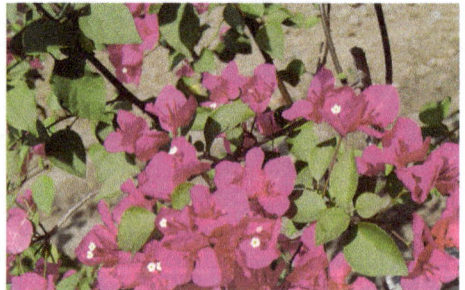

IIHR-Sonnet

*****

9

Classification of Varieties

The diversity of Bougainvillea varieties is very wide so far as vegetative and floral characters are concerned. Moreover, there are hundreds of varieties having varied growth habit, bract formation and colour besides leaf variegation. Many of them profusely bloom during main season during March to June in tropical countries. On other hand, another set is there which produce colourful bracts in winter and moderate season. Considering the wide range of variability, there is a need for classification of the varieties so that they are used in various ways as per their suitability.

A broad classification has been made on the basis of various morphological characters (vegetative and floral). This will help in categorizing the varieties and at the same time facilitate use of Bougainvilleas in proper ways in ornamental gardening.

Formation of Bracts

It is very easy to categories varieties of Bougainvilleas considering bract structure and formation. Accordingly, they are grouped into following two groups.

Single Bracted – Bracts usually form in single whorl, free, three in numbers; flower present. e.g. 'Abhimanyu', 'Blondie', 'Chitra', 'Dream', 'Dr. H.B. Singh', 'Glabra', 'Palekar', 'Shubhra', 'Shweta', 'Tetra Mrs. McClean' etc.

Multiple Bracted – Bracts usually form in multiple whorls, many in numbers and clustered; without flower. 'Archana', 'Cherry Blossom', 'Los Banos Beauty', 'Los Banos Variegata', 'Mahara', 'Mahara Variegata', 'Marietta', Roseville's Delight.

Sensation

'Thai Cherry'

Colour of the Bracts

Bract colours of Bougainvillea have lot of variation having various shades of single colour and in combination. Accordingly, the varieties have been classified in the following way.

Single Coloured

White – Bract colour white or whitish or greenish-white in pure form.eg. 'Hawaiin White', 'Himani', 'Jennifer Fernie', 'Shubhra', 'Shweta', 'Sova' etc.

Pink - Bract colour pink or shade of pink in pure form. 'Mahatma Gandhi', 'Mary Palmer', 'Poultoni', 'Poultoni Special' etc.

Red - Bract colour red or shade of red in pure form. 'Camarilo Fiesta', 'Dr. R.R. Pal', 'Flame', 'Isabel Green Smith', 'Mahara', 'Meera', 'Mrs. Butt', Mrs. R.B. Carrick', 'Palekar', 'Scarlet Glory', 'Scarlet Queen', 'Sri Durga', 'Versicolour' etc.

Purple - Bract colour purple or shade of purple in pure form. 'Formosa', 'Glabra', 'Glabra Magnifica', 'Glabra Sanderiana', 'Refulgens', 'Splendens', 'Zulu Queen' etc.

Yellow - Bract colour yellow or shade of yellow in pure form. 'Enid Lancaster', 'Golden Glow, 'Lady Mary Baring', 'Suverna', Yellow Queen'.

Orange – The shade of orange also varies from variety to variety. Representative varieties are – 'Lateritia', 'Louse Wathen', 'Margery Lloyed', 'Mrs. McClean', 'Roseville's Delight', 'Tetra Mrs. Mc Clean', 'Zakiriana' etc.

Mauve – The varieties in this category show bract colour from light to dark mauve eg., 'Dr. H.B. Singh', 'Dream', 'Enid Walker', 'President', 'Trinidad' etc.

Magenta – Representative varieties are - 'Asia', 'Gopal', 'Manohar Chandra', 'Mrs. H.C. Buck', 'Ruarka', 'Sonnet' etc.

'Refulgens'

'Splendens'

NBRI-Aruna

NBRI-Shubhra

Bi-coloured

The colour of the bracts consisted of two main colours in various combinations and patterns. The bi-cloured varieties along with their bract colour are given below.

'Begum Sikander' – Bracts having a combination of roseine purple and parchment white; rare, very restricted availability.

'Cherry Blossom' - Popular multi-bracted variety with pinkish-white-purple, mass bloomer; good availability.

'Chitra' - Excellent cultivar with large multi-coloured bracts in combination of magenta-rose and perahment white; very popular, good availability.

'Fantasy' – Bracts show two types of colour combining rosy-purple and sulphur-purple, very restricted availability.

'Mary Palmer' – Three types of bracts colour in combination of magenta and perchament white; good availability.

'Mary Palmer Special' - Exceptionally large bracts in combination of three colours – magenta, magenta-perchament white and parchment white; restricted availability.

'Odisee' – A very special kind of combination of colour comprising of light rose, parchment white and light purple in irregular blotches on large bracts, very attractive and free flowering; availability restricted.

'Thimma' – Three types of bract colour in single plant, varying from magenta-rose, parchment white and magenta-white; very attractive and hardy; easily available.

'Wajid Ali Shah' – At immature stage bracts contains irregular blotches of perchament white and roseine–purple changing to darker shade of roseine purple on attaining maturity; rare, availability is very restricted.

'Odisse'

NBRI-Begum Sikander

Leaf Variegation

Leaves of some of the varieties have unique combination of colours and pattern, very attractive. The colours vary from yellow-green, white-green, pinkish white-green, pale green-creamy. Some outstanding varieties in this category have been furnished below.

'**Archana**'- Matured leaves are having green patch at the centre and margin light yellow.

'**Arjuna**'- Leaves light grayish-green at the centre while margins are creamy-white at maturity.

'**Cinderella**'- Variegation in combination of cream and green, attractive.

'**Gangamma**'- Margin white, center green.

'**Gangaswamy**'- Centre green, white along the margin.

'**Hawaiin Beauty**'- A combination of grey-green and pale yellow in irregular blotches covering the whole leaf.

'**L.N. Birla**'- Variegation in mosaic pattern with green and cream.

'**Louise Wathen Variegata**'- Irregular variegation of grey-green at the centre and creamy-white margin.

'**Los Banos Variegata**'- Very prominent variegation; green at the centre in irregular pattern with thick white coloured margin, little wavy, very attractive.

'**Los Banos Variegata - Jayanthi**'- Variegation in mosaic pattern in various shades on green and yellow.

'**Mahara Variegata**'- Matured leaves grey-green at the centre with creamish-yellow margin in various changing pattern from one leaf to another.

'**Manohar Chandra Variegata**'- Light green centre and light yellow margin.

'**Marietta**'- Variegation in irregular fashion having grey-green at the centre and cream along the margin.

'**Mataji Agnihotri**'- Centre green with light yellow margin.

'**Mrs. Butt Variegata**'- Variegation in irregular pattern combining dark green and yellow.

'**Nirmal**'- Irregularly variegated with green centre and pale yellow margin.

'**Pallavi**'- A combination of light and pale green in various shade.

'**Parthasarthy**'- Variegation on irregular mosaic pattern with green and pale white.

'**P.V. Sane**'- Juvenile leaves bi-coloured, centre greenish in irregular pattern, outer pinkish red; matured leaves light yellow-green, centre yellowish.

'**Royal Daupline**'- Attractive combination of grey-green at the centre in irregular pattern while peripheral portion is greenish-yellow.

'**Scarlet Queen Variegata**'- Very prominent variegation; centre light grey-green in irregular pattern while rest of the portion is dull creamy-yellow.

'**Surekha**'- Light green centre and pale yellow margin.

'**Thimma**'- A combination of light yellow (centre) and green (periphery).

'**Vishakha**'- Light green at the centre and white along the margin.

'Pixie Variegata'

NBRI-Dr. A.P.J. Abdul Kalam

NBRI-Hawaiin Beauty

NBRI-Los Banos Variegata

Use

Bougainvilleas are used in various ways in ornamental gardening depending upon the requirement and suitability. It is, therefore, most important to know the right varieties for specific purpose. Accordingly, the varieties have been classified for various usages.

Arch / Pergola - 'Cherry Blossom', 'Los Banos Variegata', 'Mahara', 'Palekar', 'Roseville's Delight', 'Shubhra', 'Zulu Queen'.

Bonsai – 'Dream', 'Dr. H.B. Singh', 'Dr. R.R. Pal', 'Glabra', 'Glabra Sanderiana', 'Palekar', 'President', 'Scarlet Queen Variegata', 'Shubhra', 'Zulu Queen.'

Bush – 'Blondie', 'Dream', 'Dr. R.R. Pal', 'Glabra', 'Mary Palmer', 'Mrs. H.C. Buck', 'Red Triangle', 'Shubhra', 'Thimma', 'Zulu Queen'.

Cascade – 'Dream', 'Dr. H.B. Singh', 'Dr. R.R. Pal', 'Glabra', 'Mary Palmer', 'Palekar', 'Shubhra' etc.

Climber – 'Chitra', 'Dr. R.R. Pal', 'Lady Mary Baring', 'Mary Palmer', 'Mrs. H.C. Buck', 'Palekar', 'Partha', Shubhra', 'Splendens', 'Shubhra', 'Thimma' etc.

Espalier – 'Chitra', 'Dr.R.R.Pal', 'Dream', 'Lady Mary Baring', 'Mrs. H.C. Buck', 'Mary Palmer', 'Palekar', 'Red Triangle', 'Shbhra', 'Thimma' etc.

Grond Cover – 'Blondie', 'Dr.H.B.ingh', 'Dr. R.R.Pal, 'Glabra', 'Mrs. H.C.Buck', 'Mary Palmer', 'Palekar', 'Shubhra', 'Thimma', 'Zulu Queen' etc.

Hanging Basket – 'Aruna', 'Dream', 'Dr. H.B. Singh', 'Glabra', 'Glabra Variegata', 'Los Banos Variegata', 'Mrs. H.C. Buck', 'Palekar', 'Scarlet Queen Variegata', 'Zinna Barat', 'Zulu Queen'.

Road side and on dividers – 'Blondie', 'Dream', 'Dr. R.R. Pal', 'Glabra', 'Mary Palmer', 'Mrs. H.C. Buck', 'Palekar', 'Red Triangle', 'Shubhra', 'Thimma', 'Zulu Queen'.

Standard - 'Chitra', 'Lady Mary Baring', 'Los Banos Variegata', 'Mary Palmer Special', 'Palekar', 'Poultoni Special', 'Shubhra', 'Tetra Mrs. McClean', 'Thimma', 'Zulu Queen.

Hedge – 'Dream', 'Dr. R.R. Pal', 'Thimma', 'Mrs. H.C. Buck', 'Mrs. McClean', 'Partha', 'Tomato Red'.

Pot Plant – 'Chery Blossom', 'Chitra', 'Dr. R.R. Pal', 'Mrs. H.C. Buck', 'Los Banos Variegata', 'Mary Palmer', 'Palekar', 'Poultoni Special', 'Shubhra', 'Thimma' etc.

Based on Blooming Season

Development of bracts of Bougainvilleas (blooming) takes place in flashes in various seasons. Therefore, it is important to know the blooming pattern of the varieties for judicious use in landscaping.

Round the Year – These varieties are recurrent bloomer and produce bracts around the year with varied intensity. 'Aruna', 'Dr. H.B. Singh', 'Dr. P.V. Sane', 'Dream', 'Palekar', 'Mary Palmer Special', 'Mrs. H.C. Buck', 'Red Triangle' etc.

Winter Blooming – These varieties produce bracts during winter months (October to January) over and above the normal blooming period. 'Begum Sikandar', 'Chitra', 'Los Banos Variegata', 'Mahara', 'Roseville's Delight', 'Splendens' 'Wajid Ali Shah',

 Summer Blooming – These varieties profusely bloom during summer months (March to June) but remained out of bloom during other season. 'Arjuna', 'Dr. H.B. Singh, Isabel Green Smith', 'Mary Palmer Special', 'Partha', 'Shubhra', 'Tetra Mrs. McClean', 'Thimma' etc.

*****

10

Use in Ornamental Gardening

Bougainvilleas create colourful effect on the landscape by their attractive bracts in various colours. No other plants can alter the face of the gardens in such a magnificent way. Considering the impact, Bougainvilleas are used in the ornamental gardening in large scale and in various ways. Their varied adaptability and growth habit facilitate usages in different styles.

Some of the most prominent ways of using Bougainvilleas in ornamental gardening have been furnished along with varieties for specific use.

Arch and Pergola – These are generally placed at the gate / entrance and at suitable points. Arch and Pergola are common structure of every large parks and gardens for training creepers. Bougainvilleas are ideal for this purpose and suitable varieties are – 'Cherry Blossom', 'Dr. R.R.Pal', 'Los Banos Variegata', 'Mahara', 'Mary Palmer' 'Palekar', 'Roseville's Delight', 'Shubhra', 'Zulu Queen'.

Bonsai – Bougainvilleas can easily be trained as bonsai by repeated pruning and pinching in different styles by planting them in decorative trays. These specimens flower in March-April, and are very pretty to see under leafless condition. Flowering Bonsai specimens are excellent items for display in lighted places of the garden and interiors. Varieties suitable for bonsai purpose – 'Dr. H.B. Singh', 'Dream', 'Dr. R.R. Pal', 'Glabra', 'Joe', 'Scarlet Queen Variegata', 'Thimma', 'Palekar', 'Isabel Green Smith', 'Zulu Queen' etc.

Bush – Bougainvilleas form good bush when planted in ground. Proper designing by planting in series combining various colour of bracts exhibits an attractive display. Huge mass full of flowers, is an effective way displaying mass effect of bract colours. Recommended varieties are – 'Blondie', 'Chitra', 'Dream', 'Dr. R.R Pal', 'Golden Glow', 'Glabra', 'Mrs. H.C. Buck', 'Red Triangle', 'Thimma', 'Zulu Queen' etc.

Cascade – In this training, Bougainvilleas grown in a pot and finally the shoots are allowed to trail over the framed structure having the shape of a bettle leaf. Thereafter, pruning and pinching are carried out to retain the shape of the cascade using the frame. It is an unconventional type of training and need experience for doing this successfully. Varieties suitable for this purpose are – 'Aruna', 'Dr. H.B. Singh', 'Dr. R.R. Pal', 'Dream' 'Glabra', 'Isabel green Smith', 'Mary Palmer', 'Nirmal', 'Palekar', 'Scarlet Queen Variegata', etc.

Climbers – Bougainvilleas as climber produce a magnificent effect if properly placed at the back drop of building, pillars and gate. Proper training, pruning and selection of right varieties are the key points for creating effective display. 'Dr.R.R. Pal', 'Lady Mary Baring', 'Louse Wathen', 'Mary Palmer', 'Roseville's Delight', 'Royal Daupline', 'Shubhra', 'Thimma' can be used for this purpose.

Espalier – This is a special kind of training for beautification of the walls. Selected varieties are planted near the wall and main stem is allowed to grow without any side branches up to 75-90 cm. Thereafter, side branches are allowed in opposite direction. Stem and branches are fixed on the wall by wire and nails. By repeating this process, a uniform style is created which gives an elegant look when the plant is in bloom. This type of training requires experience and skill on training pruning. Suitable varieties are – 'Aruna', 'B.T. Red', 'Chitra', 'Dr. R.R. Pal', 'Dream', 'Lady Mary Baring', 'Mrs. H.C. Buck', 'Mary Palmer', 'Palekar', 'Red Triangle', 'Shubhra', 'Thimma' etc.

Ground Cover – Bougainvilleas are also grown as ground cover by selecting those varieties having prostrate habit. Uneven, rocky and sloppy spots in the garden can be turned into pleasing spot by planting selected bougainvillea varieties. Suitable varieties are – 'Aruna', 'Blondie', 'Dr.H.B.Singh', 'Dream', 'Shubhra', 'Palekar', ' Glabra' etc.

Hanging Basket – The idea of growing *Bougainvillea*s in hanging baskets is relative new. Those varieties which have drooping branches should be selected. Repeated pruning of vertical growths and allowing only lateral braches produce a hanging effect. Varieties suitable for this purpose are – 'Aruna', 'Dr. H.B. Singh', 'Glabra Variegata', 'Gillian Greensmith', 'Isabel Greensmith', 'Palekar', 'Scarlet Queen Variegata', 'Zinna Barat' etc.

Hedge – Bougainvilleas are very suitable for development of hedge. Due to the presence of spines, a protective hedge along the boundary wall can be developed by using different varieties. Moreover, selection of varieties having variegated leaves makes a useful display of hedge. Varieties recommended for developing hedge are – 'Dream', 'Dr. R.R. Pal', 'Mrs. H.C. Buck', 'Mrs. Mc Clean', 'Los Banos variegata', 'Partha', 'Scarlet Queen Variegata', 'Thimma', 'Tomato Red' etc.

Slopes and Mounds – There are some slopes or mounds in every parks and garden which are very difficult to beautify effectively with other plants. This problem can easily be solved by planting selected Bougainvillea varieties which can beautify those areas. Recommended varieties are – 'Aruna', 'Dream', 'Palekar', 'Shubhra', 'Mrs. H. C. buck', 'Mary Palmer', 'Dr.R.R.Pal'

Standard –A unique type of training on a G.I. pipe stand (1.5 m from ground level with m.s. round and spokes having 1m diameter at the top). The plant is allowed initially, to grow on a single stem up to the top of the stand. Afterwards, branching is allowed into different direction so that ultimately the plant takes a shape of an umbrella. It requires constant monitoring and maintenance by clipping and pruning to keep the plants in proper shape.

Different varieties are selected for the development of 'Standards' and planted by the side of passage or inner road in straight line. When in full bloom, these standards create a scenic beauty. Varieties recommended for this purpose are – 'Aruna', 'Chitra', 'Lady Mary

Baring', 'Los Banos Variegata, 'Mary Palmer Special', 'Palekar', 'Poultoni Special', 'Shubhra', 'Tetra Mrs. McClean', 'Thimma', 'Zulu Queen' etc.

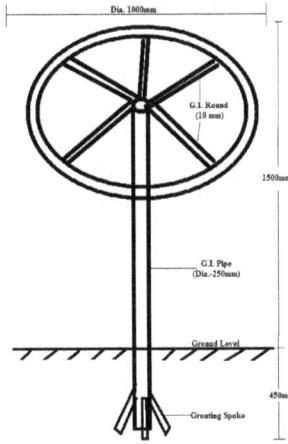

Design of the structure of 'Standard'

Training on tree or stump – This is a simple technique of training Bougainvilleas over the trees or stump, as climb well over the support. These give a magnificent effect and can be used in roadside decoration. Varieties recommended when in bloom, for this purpose are – 'Dream', 'Dr. R.R. Pal', 'Glabra', 'Mrs. H.C. Buck', 'Palekar', 'Shubhra', 'Thimma', etc.

Other Usages of Bougainvilleas

In addition to ornamental use, Bougainvilleas are used in various other ways as a plant for avenue decoration and pollution abatement.

Planting along the Road - Bougainvilleas are relatively tolerant to pollutions. Therefore, they are recommended for plantation as companion plant along with trees planted by the roadside. They combine well and produce colourful effect. Varieties having hardy and vigorous growth habit should be selected.

Central Verge - In the cities and towns, these are often found neglected and devoid of any planting. It is recommended that these areas should be well utilized by planting dwarf trees and shrubs. Planting may be done either in single or double row depending upon the space available. Since these plants are more close to the automobile exhaust, their capacity for pollution tolerance should be considered before selection. Bougainvilleas being pollution tolerant are recommended for plantation in the central verge. Varieties suitable for this purpose are 'Chitra', 'Dr. H.B Singh', 'Dr. P.V. Sane', 'Glabra', 'H.C. Buck', 'Lady Mary Baring', 'Mahara', 'Mary Palmer Special', 'Partha', 'Palekar' 'Shubhra', 'Thimma' etc.

Traffic Island - These vary in shape and size from square, triangle to round. These islands should be properly planted with the dwarf trees, shrubs and ground covers recommended for planting along the road and central verge which will contribute effectively in mitigating the air pollution. These areas are ideal for plantation of Bougainvilleas in combination with other ornamental plants. Following varieties are recommended for plantation in Traffic Island – 'Dream', 'Dr. H.B. Singh', 'Glabra', 'Mrs. H.C. Buck', 'Thimma', 'Shubhra', 'Dr. R.R Pal' etc.

Greenbelt - In urban areas certain areas are earmarked for the development of greenbelts. In greenbelt, Bougainvilleas are planted in combination with trees to add colour to the landscape and creates aesthetic effect. Varieties having vigorous growth habit are suitable for this purpose.

Window Sill - In cities, usually multi-storeyed buildings of corporate offices, hotels, residential complexes have well designed window sills for plantation exclusively for beautification purpose. Bougainvilleas are best for plantation in window sill. The hanging effect produced together with colourful bracts creates a very attractive scene aesthetically. Varieties having variegated foliage are also ideal for their colourful foliage in addition to bracts. The recommendable varieties are - 'Aruna', 'Chita', 'Dr. H.B. Singh', 'Dr. R.R. Pal', 'Dream', 'Glabra', 'Mary Palmer', 'Palekar', 'Royal Daupline', 'Shubhra', 'Scarlet Queen Variegata'.

Artistic arrangement of bract colour

Artistic Training

Blazing colour of bracts ('Splendens')

BougainvilleaGarden

Cascading Effect

Climbing effect on tree

Mass display in plant house

On boundary wall of a home garde**

Picturesque effect of Shubhra

Plantation along the road

Plantation in combination with annuals

Potted plants in group

Training as csacade

Training as wall in containers

Training in Coloumn

Training on boundary wall

Wall Cover

11
List of Available Varieties

A comprehensive list of some prominent available varieties along with their parentage, growth habit, morphological characters (floral and vegetative parameters) has been provided in this table. This will serve as a tool for identification and distinguishing one variety from the other.

S. No.	Name of the Varieties	Description
1	'Abhimanyu'	**Species:** *Bougainvillea peruviana* **Parent:** Bud sport of 'Arjuna' **Habit:** Medium, drooping, growth restricted. **Vegetative Characters:** Stem young greenish-coppery; Leaves variegated (margin creamish), leaf blade - 5.8 x 3.0 cm, elliptic, Green group 137 C, Fan-3; Thorn 1.1 cm long, straight. **Flower/Bract Characters:** Flowering profuse, at the end of the branches; Bracts 4.0 x 2.8 cm, elliptic, Red-purple group 72 B, Fan-2; star 0.8 cm dia., non-persistent; Star 0.8 cm dia., Yellow group 2 D, Fan -1.
2	'Abraham Kavoor'	**Species:** *Bougainvillea x buttiana* **Parent:** Unknown **Habit:** Tall, vigorous growth. **Vegetative Characters:** Young stem greenish-coppery; Leaf blade - 6.6 x 5.1 cm, ovate, Green group 137 C, Fan-3; Thorn 0.8 cm long, straight. **Flower/Bract Characters:** Flowering profuse, 1/2 of the branches; Bracts 4.3 x 2.8 cm, ovate, Yellow-orange group 20 B, Fan-1, non-persistent; Star 0.6 cm dia., Green-yellow group 1 D, Fan-1.

3   'Aida'

Species: *Bougainvillea spectabilis*

Parent: Hybrid seedling of *Bougainvillea spectabilis*

Habit: Tall, growth vigorous.

Vegetative Characters: Stem young coppery; Leaf blade 7.0 x 4.6 cm, ovate, Green group 137 C, Fan-3; Thorn 1.1 cm long, straight.

Flower/Bract Characters: Flowering profuse, at the end of the branches in compact masses; Bracts 4.3 x 2.8 cm, elliptic, Red-purple group 71 B, Fan-2, non-persistent; Star 0.6 cm dia., Green-yellow group 1 D, Fan -1.

4   **'Aida Variegata'**

Species: *Bougainvillea spectabilis*

Parent: Bud Sport of 'Aida'

Habit: Tall, growth vigorous.

Vegetative Characters: Stem young coppery; Leaf blade 7.0 x 4.6 cm, ovate, Green group 137 A, Fan-3, leaves variegated irregularly (yellowish green speckled); Thorn 1.1 cm long, straight.

Flower/Bract Characters: Flowering profuse, at the end of the branches in compact masses; Bracts 4.3 x 2.8 cm, elliptic, Red-purple group 64 A, Fan-2, non-persistent; Star 0.6 cm dia., Yellow group 5 D, Fan-1.

5   **'Alick Lancaster'(' Lilac Queen')**

Species: *Bougainvillea x buttiana*

Parent: Natural Mutation

Habit: Dwarf, growth restricted.

Vegetative Characters: Stem young greenish coppery; Leaf blade 6.6 x 3.1 cm, elliptic, Yellow-green group 144 A, Fan-3; Thorn 1.0 cm long, straight.

Flower/Bract Characters: Flowering profuse, Bracts borne all along the branches; Bracts 4.1 x 2.3 cm, elliptic, Red-purple group 70 D, Fan-2, non-persistent; Star 0.8 cm dia., Green-yellow group 1 C, Fan-1.

6   **'Allizon Davy'**

Species: *Bougainvillea buttiana*

Parent: Unknown

Habit: Tall, growth intermediate.

Vegetative Characters: Stem young coppery; Leaf blade 5.8 x 3.3 cm, elliptic, variegated, Green group 137 C, Fan-3; Thorn 1.1 cm long, straight.

Flower/Bract Characters: Flowering sparse, at the end of the branches; Bracts 3.3 x 2.4 cm, ovate, Red-purple group 63 B, Fan-2, non-persistent; Star 0.6 cm dia., Green-yellow group 1 D, Fan-1.

7   **'Annabella'**

Species: *Bougainvillea glabra*

Parent: Hybrid seedling

Habit: Medium, growth intermediate.

Vegetative Characters: Stem young green; Leaf blade 10.0 x 3.6 cm, elliptic, Green group 137 B, Fan-3; Thorn 1.5 cm long, slightly curved at tip.

Flower/Bract Characters: Flowering profuse, all along the branches; Bracts 3.4 x 1.6 cm, ovate, Red-purple group 75 D, Fan-2, persistent; Star 0.8 cm dia., Green-yellow group 1 C, Fan-1.

8 **'Arjuna'**

Species: *Bougainvillea peruviana*

Parent: Bud sport of 'Partha'

Habit: Medium, drooping growth restricted.

Vegetative Characters: Stem young green; Leaves variegated (creamish white margin), leaf blade 8.1 x 3.7 cm, elliptic, Green group 138 B, Fan-3; Thorn 1.2 cm long, straight.

Flower/Bract Characters: Flowering profuse, at the end of the branches in compact masses; Bracts 4.0 x 2.6 cm, elliptic, Purple group 77 B, Fan-2, non-persistent; Star 0.8 cm dia., Yellow group 2 D, Fan-1.

9 **'Aruna'**

Species: *Bougainvillea peruviana*

Parent: Bud sport of 'Partha'

Habit: Medium, drooping, growth vigorous.

Vegetative Characters: Stem young greenish-coppery; Leaf blade 6.1 x 3.2 cm, elliptic, Green group 137 A, Fan-3, curved; Thorn straight, 1.2 cm long.

Flower/Bract Characters: Flowering profuse, 1/2 along the branches, Bracts 2.6 x 1.6 cm, elliptic, Orange Red group 32 C, Fan-1, non-persistent; Star 0.7 cm dia., Yellow group 2 D, Fan-1.

10 **'Asia'**

Species: *Bougainvillea x buttiana*

Parent: Hybrid seedling of *Bougainvillea x buttiana*

Habit: Dwarf, growth restricted.

Vegetative Characters: Stem young greenish-coppery; Leaf blade 6.7 x 3.8 cm, ovate, Green group 137 A, Fan-3; Thorn 0.9 cm long, straight.

Flower/Bract Characters: Flowering profuse, all along the branches in compact masses; Bracts 3.1 x 2.2 cm, ovate, Red-purple group 72 B, Fan-2, non-persistent; Star 0.6 cm dia., Green-yellow group 1D, Fan-1.

11 **'B. T. Red'**

Species: *Bougainvillea x buttiana*

Parent: Unknown

Habit: Medium, growth intermediate.

Vegetative Characters: Stem young green; Leaf blade 5.1 x 3.2 cm, ovate, Green group 137 C, Fan-3; Thorn 0.9 cm long, straight.

Flower/Bract Characters: Flowering sparse, at the end of branch; bracts 3.1 x 2.2 cm, ovate, Red-purple group 72 A, Fan-2, non-persistent; Star 0.5cm dia., Yellow group 2 D, Fan-1.

12 **'Barbara Karst'**

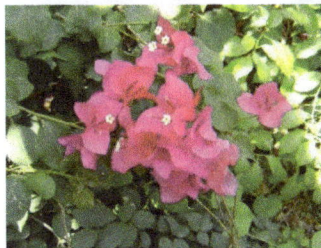

Species: *Bougainvillea x buttiana* and *B. glabra*

Parent: Hybrid seedling of 'Crimson Lake' x 'Sanderiana'

Habit: Tall, growth vigorous.

Vegetative Characters: Stem young coppery; leaf blade 6.3 x 3.8 cm, ovate, Green group 137 A, Fan-3; Thorn 1.8 cm long, straight.

Flower/Bract Characters: Flowering profuse, at the end of the branches; Bracts 3.7 x 2.6 cm, ovate, Red-purple group 63 A, Fan-2, non-persistent; Star 0.6cm dia., Green-yellow group 1 D, Fan-1.

13 'Begum Sikander'

Species: *Bougainvillea peruviana* and *B.glabra*

Parent: Hybrid seedling of 'Dr. B.P. Pal' x 'Jennifer Fernie'

Habit: Medium, growth restricted.

Vegetative Characters: Stem young greenish-coppery; Leaf blade 10.3 x 5.1 cm, ovate to elliptic, Green group 137 A, Fan-3; Thorn 2 cm long, straight.

Flower/Bract Characters: Flowering profuse, all along the branches; Bracts 4.5 x 3.0 cm, ovate, double colour - Red-purple group 65 B, Fan-2 and White group 155 D, Fan-4, non-persistent; Star 1.0 cm dia., Green-yellow group 1C, Fan-1.

14 'Blondie'

Species: *Bougainvillea x buttiana*

Parent: Hybrid seedling of *Bougainvillea x buttiana*

Habit: Tall, growth vigorous.

Vegetative Characters: Stem young green; leaf blade 6.7 x 3.5 cm, elliptic, Green group 137 A, Fan-3; Thorn 1.5 cm long, slightly curved.

Flower/Bract Characters: Profuse flowering, at the end of the branches; Bracts 3.9 x 2.8 cm, Red-purple group 65 B, Fan-2, ovate, non-persistent; Star 0.6cm dia., Yellow group 2D, Fan-1.

15 'Brasiliensis'

Species: *Bougainvillea spectabilis*

Parent: Unknown

Habit: Tall, growth vigorous.

Vegetative Characters: Stem young coppery; Leaf blade 4.7 x 3.4 cm, ovate, Green group 137 C, Fan-3; Thorn 0.9 cm long, straight.

Flower/Bract Characters: Flowering profuse,1/2 along the branches; Bracts 3.8 x 3.3 cm, ovate, Red-purple group 70 A, Fan-2, non-persistent; Star 0.7cm dia., Green-yellow group 1C, Fan-1.

16 'Brilliant'

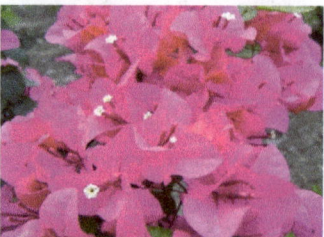

Species: *Bougainvillea x buttiana*

Parent: Seedling of *Bougainvillea x buttiana*

Habit: Tall, growth vigorous.

Vegetative Characters: Stem young greenish-coppery; Leaf blade 4.5 x 3.3 cm, ovate, Yellow-green group 144A, Fan-3; Thorn 0.8 cm long, straight.

Flower/Bract Characters: Flowering profuse, at the end of the branches; Bracts 4.0 x 3.3 cm, ovate, Red-purple group 70 B, Fan-2, non-persistent; Star 0.7cm dia., Yellow group , Fan-1.

17 'Camarillo Fiesta'

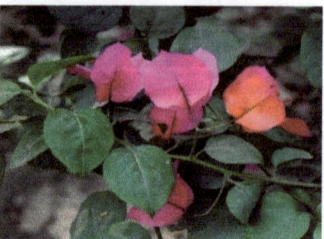

Species: *Bougainvillea x buttiana*

Parent: Unknown

Habit: Medium, growth vigorous.

Vegetative Characters: Stem young coppery; Leaf blade 6.7 x 4.5 cm, ovate, Green group 137 C, Fan-3; Thorn 0.8 cm dia., curved.

Flower/Bract Characters: Flowering medium, 1/2 along the branches; Bract 3.5 x 2.7 cm, ovate, Orange group 25 B, Fan-1 to Red-purple group 70 C, Fan-2, non-persistent; Star 0.8 cm dia., Yellow group 2 D, Fan-1.

18 **'Carloton Corea'**

Species: *Bougainvillea x buttiana*

Parent: Unknown

Habit: Medium, growth vigorous.

Vegetative Characters: Stem young coppery; Leaf blade 8.3 x 5.5 cm, ovate, Green group 137 A, Fan-3, curved; Thorn 0.9 cm long, straight.

Flower/Bract Characters: Flowering medium, 1/3 along the branch; Bract 4.4 x 3.1 cm, broadly ovate, Red-purple group 71 C, Fan-2, persistent; Star 0.8 cm dia., Yellow group 2 D, Fan-1.

19 **'Cascade'**

Species: *Bougainvillea x buttiana*

Parent: Hybrid seedling

Habit: Tall, growth vigorous.

Vegetative Characters: Stem young coppery; Leaf blade 6 x 3.7 cm, ovate, Green group 137 C, Fan-3, curve; Thorn 0.8 cm long; straight.

Flower/Bract Characters: Flowering sparse, 1/3 along the branch; Bract 3.5 x 2.2 cm, elliptic, Red-purple group 61 A, Fan-2, non-persistent; Star 0.7 cm dia., Yellow group 3 D, Fan-1.

20 **'Charles William'**

Species: *Bougainvillea peruviana*

Parent: Unknown

Habit: Tall, growth vigorous.

Vegetative Characters: Stem young greenish-coppery; Leaf blade 6.8 x 3.9 cm, ovate, Green group 137 C, Fan-3; Thorn 1.1 cm long, straight.

Flower/Bract Characters: Flowering profuse, at the end of the branches; Bract 3.7 x 2.8 cm, ovate, Red-purple group 70 A, Fan-2, non-persistent; Star 0.6 cm dia., Yellow group 2 D, Fan-1.

21 **'Cherry Blossom'**

Species: *Bougainvillea x buttiana*

Parent: Bud sport of 'Los Banos Beauty'

Habit: Tall, growth vigorous.

Vegetative Characters: Stem young greenish-coppery; Leaf blade 5.8 x 3.9 cm, ovate, Green group 137 C, Fan-3; Thorn 1.0 cm long, slightly curved.

Flower/Bract Characters: Flowering profuse, at the end of the branches; Multibracted, 2.0 x 1.2 cm, elliptic to ovate, Red group 55 D, Fan-1, persistent; Flower absent.

22 **'Chinese Cracker'**

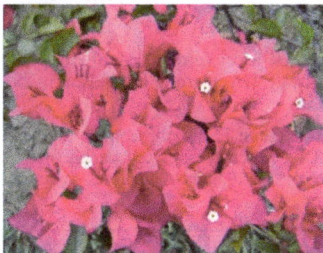

Species: *Bougainvillea x buttiana*

Parent: Unknown

Habit: Tall, growth intermediate.

Vegetative Characters: Stem young greenish-coppery; Leaf blade 5.4 x 3.4 cm, broadly ovate, Green group 137 C, Fan-3; Thorn 0.8 cm long, slightly curved.

Flower/Bract Characters: Flowering profuse, at the end of branches; Bracts 3.7 x 2.8 cm, ovate, Red-purple group 60 B, Fan-2, non-persistent; Star 0.6 dia., Yellow group 2 D, Fan-1.

23   **'Chitra'**

Species: *Bougainvillea x buttiana* and *B. peruviana*

Parent: Hybrid Seedling of 'Tetra Mrs. McClean' x 'Dr. B. P. Pal'

Habit: Tall, growth vigorous.

Vegetative Characters: Stem young coppery; Leaf blade 7.9 x 6.0 cm, broadly ovate, Green group 137 A, Fan-3; Thorn 1.4 cm long, slightly curved.

Flower/Bract Characters: Flowering medium, at the end of the branch, Bracts 3.6 x 3.1 cm, broadly ovate, double coloured - Red-purple group 70 C, Fan-2 and Red group 56 D, Fan-1 non-persistent; Star 0.9 cm dia., Yellow group 2 D, Fan-1.

24   **'Cleopatra'**

Species: *Bougainvillea peruviana*

Parent: *Bougainvillea peruviana*

Habit: Tall, growth vigorous.

Vegetative Characters: Stem young greenish-coppery; Leaf blade 6.1 x 3.8 cm, ovate, Green group 137 D, Fan-3; Thorn 1.2 cm long, straight.

Flower/Bract Characters: Flowering medium, at the end of the branch; Bract 3.6 x 2.6 cm, ovate, Red-purple group 71 B, Fan-2, non-persistent; Star 0.6 cm dia., Yellow group 3 D, Fan-1.

25   **'Crispa'**

Species: *Bougainvillea x buttiana*

Parent: Unknown

Habit: Dwarf, growth restricted.

Vegetative Characters: Stem young green; Young leaves variegated, leaf blade 4.6 x 3.4 cm, broadly ovate, Green group 139 A, Fan-3, margin crisped heavily; Thorn 0.6 cm long, straight.

Flower/Bract Characters: Flowering sparse, 1/3 of the branch; Bracts 3.2 x 2.4 cm, ovate, Red-purple group 63 B, Fan-2, twisted, non-persistent; Star 0.5 cm dia., Green-yellow group 1 C, Fan-1.

26   **'Double Delight'**

Species: *Bougainvillea peruviana*

Parent: Unknown

Habit: Tall, growth vigorous.

Vegetative Characters: Stem young greenish coppery; Leaf blade 7.2 x 4.8 cm, broadly ovate, apex acuminate, variegated, Green group 137 A, Fan-3, Greyed-green group 191 A, Fan-4, Yellow group 11 B, Fan-1, ; Thorn 1.5 cm in long, slightly curved.

Flower/Bract Characters: Flowering profuse, all along the branches; Bract 4.5 x 3.7 cm, broadly ovate, double colour - Red-purple group 64 C, Fan-2, Red group 56 D, Fan-1 (Whitish-pink), non-persistent; Star 0.8 in dia., Green-yellow group 1 D, Fan-1.

27   **'Dog Star'**

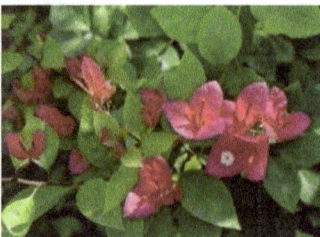

Species: *Bougainvillea spectabilis* (*Bougainvillea x buttiana*)

Parent: Hybrid seedling of *Bougainvillea spectabilis*

Habit: Medium, growth vigorous.

Vegetative Characters: Stem young greenish-coppery; Leaf blade 6.1 x 3.3 cm, ovate, Green group 137 B, Fan-3; Thorn1.0 cm long, straight.

Flower/Bract Characters: Flowering sparse, all along the branches; Bracts 3.8 x 2.6 cm, ovate, Red-purple group 71 B, Fan-2, non-persistent; Star 0.8 cm dia., Green-yellow group 1 D, Fan-1.

28 **'Dr. H.B. Singh'**

Species: *Bougainvillea glabra*

Parent: Hybrid seedling of 'Trinidad' and 'Formosa'

Habit: Dwarf, drooping, growth restricted.

Vegetative Characters: Stem young green; Leaf blade 6.8 x 2.7 cm, elliptic, Green group 137 C, Fan-3; Thorn 1.9 cm long, curved.

Flower/Bract Characters: Flowering profuse, all along the branches; Bracts 3.3 x 2.5 cm, elliptic, Purple group 75 A, Fan-2, non-persistent; Star 0.9 cm dia., Green-yellow group 1 C, Fan-1.

29 **'Dr. P.V. Sane'**

Species: *Bougainvillea x buttiana*

Parent: Mutant of 'Dr. R. R. Pal'

Habit: Tall, growth vigorous.

Vegetative Characters: Stem young green; Leaves variegated (yellowish green), leaf blade 8.7 x 5.6 cm, ovate, Green group 137 C, Fan-3; Thorn 1.3 cm long, curved.

Flower/Bract Characters: Flowering medium, 1/3 of the branches; Bracts 3.9 x 2.7 cm, ovate, Red-purple group 71 B, Fan-2, non-persistent; Star 0.9 cm dia., Green-yellow group 1 D, Fan-1.

30 **'Dr. R.R. Pal'**

Species: *Bougainvillea x buttiana*

Parent: Hybrid seedling of 'Thomasii' (*B. x glabra*) X 'Louise Wathen' (*B. x buttiana*)

Habit: Tall, growth vigorous.

Vegetative Characters: Stem young greenish-coppery; Leaf blade 8.8 x 5.7 cm, ovate, Green group 137 B, Fan-3; Thorn cm long, straight.

Flower/Bract Characters: Flowering profuse, free; Bracts 3.6 x 2.4 cm, ovate, Red-purple group 61 B, Fan-2, non-persistent; Star 0.6 cm dia., Green-yellow group 1 D, Fan-1.

31 **'Dr. Rao'**

Species: *Bougainvillea x buttiana*

Parent: Bud sport 'Mrs. Butt'

Habit: Medium, growth vigorous.

Vegetative Characters: Stem young coppery; Leaves variegated (whitish yellow along margin), leaf blade 6.4 x 4.4 cm, elliptic, Green group 138 A, Fan-3; Thorn 0.8 cm long, straight.

Flower/Bract Characters: Flowering medium, all along the branches; Bracts 3.3 x 2.5 cm, ovate, Red-purple group 71 C, Fan-2, non-persistent; Star 0.5 cm dia., Red group 37 C, Fan-1.

32 **'Dream'**

Species: *Bougainvillea glabra*

Parent: Seedling of *Bougainvillea glabra*

Habit: Medium, drooping, growth intermediate

Vegetative Characters: Stem young green; Leaf blade 6.9 x 4.6 cm, elliptic, Yellow-green group 144 A, Fan-3; Thorn 1.0 cm long, curved

Flower/Bract Characters: Flowering profuse, all along the branches; Bracts 3.6 x 2.2 cm, elliptic, Red-purple group 69 B, Fan-2, persistent; Star 0.7 cm dia., Green-yellow group 1 C, Fan-1

33 'Dwarf Gem'

Species: *Bougainvillea glabra*

Parent: Seedling of *Bougainvillea glabra*

Habit: Tall, growth vigorous.

Vegetative Characters: Stem young green; Leaf blade 7.9 x 4.4 cm, ovate, Green group 137 D, Fan-3; Thorn 1.7 cm long, curved.

Flower/Bract Characters: Flowering profuse, all along the branches; Bracts 3.5 x 2.7 cm, ovate, Red-purple group 70 B, Fan-2, non-persistent; Star 0.7 cm dia., Yellow group 3 D, Fan-1.

34 'Elizabeth'

Species: *Bougainvillea spectabilis*

Parent: Seedling of *Bougainvillea spectabilis*

Habit: Tall, growth vigorous.

Vegetative Characters: Stem young coppery; Leaf blade 7.9 x 4.8 cm, ovate, Green group 137 B, Fan-3; Thorn 1.6 cm long, straight.

Flower/Bract Characters: Flowering profuse, at the end of the branches; Bracts 3.6 x 2.7 cm, ovate, Red-purple group 71 A, Fan-2, non-persistent; Star 0.6 cm dia., Yellow group 2 C, Fan-1.

35 'Elizabeth Agnus'

Species: *Bougainvillea spectabilis*

Parent: Seedling of *Bougainvillea spectabilis*

Habit: Tall, growth vigorous.

Vegetative Characters: Stem young coppery; Leaf blade 5.8 x 4.1 cm, ovate, Green group 137 A, Fan-3; Thorn 1.4 cm long, straight.

Flower/Bract Characters: Flowering profuse, at the end of the branches; Bracts 3.6 x 2.2 cm, ovate, Red-purple group 72 C, Fan-2, non-persistent; Star 0.6 cm dia., Red group 50 D, Fan-1.

36 'Ena Heyneker'

Species: *Bougainvillea x buttiana*

Parent: Unknown

Habit: Medium, growth intermediate.

Vegetative Characters: Stem young greenish-coppery; Leaf blade 5.9 x 3.0cm, ovate, Green group 137 A, Fan-3; Thorn 2.0 cm long, straight.

Flower/Bract Characters: Flowering medium, all along the branches; Bracts 3.0 x 2.6 cm, ovate, Red-purple group 70 B, Fan-2, non-persistent; Star 0.5 cm dia., Green-yellow group 1 D, Fan-1.

37 'Enid Lancaster'

Species: *Bougainvillea x buttiana*

Parent: Bud sport of 'Louise Wathen'

Habit: Tall, growth vigorous.

Vegetative Characters: Stem young coppery; Leaf blade 6.1 x 4.1 cm, ovate, Green group 144 A, Fan-3; Thorn 0.9 cm long, slightly curved.

Flower/Bract Characters: Flowering profuse, at the end of the branches; Bracts 3.5 x 2.7 cm, ovate, Orange group 26 B, Fan-1, twisted, non-persistent; Star not prominent.

38 **'Enid Walker'**

Species: *Bougainvillea glabra*

Parent: Unknown

Habit: Medium, growth moderate.

Vegetative Characters: Stem young coppery; Leaf blade 7.6 x 3.8 cm, ovate, Green group 137 C, Fan-3; Thorn 0.9 cm long, straight.

Flower/Bract Characters: Flowering profuse, at the end of the branches; Bracts 3.0 x 1.7 cm, ovate, Purple group 75 B, Fan-2, non-persistent; Star 0.6 cm dia., Green-yellow group 1 C, Fan-1.

39 **'Easter Parade'**

Species: *Bougainvillea glabra*

Parent: Bud sport of 'Sanderiana'

Habit: Medium, growth moderate.

Vegetative Characters: Stem young green; Leaf blade 6.8 x 3.7 cm, elliptic, Green group 137 A, Fan-3; Thorn 1.0 cm long, curved.

Flower/Bract Characters: Flowering profuse, all along the branches; Bracts 3.7 x 2.6 cm, ovate, Red-purple group 72A, Fan-2, persistent; Star 0.6 cm dia., Green-yellowgroup 1 D, Fan-1.

40 **'Fantasy'**

Species: *Bougainvillea peruviana*

Parent: Bud sport of 'Princess Margaret Rose'

Habit: Tall, growth moderately vigorous.

Vegetative Characters: Stem young coppery; Leaf blade 7.8 x 5.1 cm, ovate, Green group 137 B, Fan-3; Thorn 1.4 cm long, straight.

Flower/Bract Characters: Flowering profuse, at the end of the branches; Bracts 4.0 x 3.1 cm, ovate, Red-purple group 72 B, Fan-2, non-persistent; Star 0.6 cm dia., Green-yellow group 1 C, Fan-1.

41 **'Feathery Fantasy'**

Species: *Bougainvillea x buttiana*

Parent: Unknown

Habit: Dwarf, growth restricted.

Vegetative Characters: Stem young greenish-coppery; Leaf blade 6.2 x 2.3 cm, elliptic, variegated (yellow at margin), Green group 137 A, Fan-3; Thorn 1.0 cm long, straight.

Flower/Bract Characters: Flowering sparse, at the end of branches; Bracts 3.4 x 1.1 cm, elliptic, Red-purple group 67 A, Fan-2, non-persistent; Star 0.5 cm dia., Green-yellow group 1 C, Fan-1.

42   **Feathery Fantasy (Double Coloured)**

Species: *Bougainvillea x buttiana*

Parent: Unknown

Habit: Dwarf, growth restricted.

Vegetative Characters: Stem young greenish-coppery; Leaf blade 6.0 x 2.1 cm, abnormal elliptic, variegated, Green group 137 C, Fan-3, Yellow group 11 B, Fan-1; Thorn 1.0 cm long, straight.

Flower/Bract Characters: Flowering sparse, at the end of branches; Bracts 3.4 x 1.1 cm, abnormal elliptic, double colour - Red-purple group 68 B, Fan-2, White group 155 B, Fan-4, non-persistent; Star 0.5 cm dia., Green-yellow group 1 C, Fan-1.

43   **'Filoman'**

Species: *Bougainvillea spectabilis*

Parent: Seedling of *Bougainvillea spectabilis*

Habit: Tall, growth vigorous.

Vegetative Characters: Stem young greenish-coppery; Leaf blade 6.8 x 4.1 cm, ovate, Green group 137 A, Fan-3; Thorn 1.1 cm long, straight.

Flower/Bract Characters: Flowering profuse, at the end of the branches; Bracts 3.6 x 2.6 cm, ovate, Red-purple group 72 B, Fan-2, non-persistent; Star 0.6 cm dia., Green-yellow group 1 D, Fan-1.

44   **'Flame'**

Species: *Bougainvillea spectabilis*

Parent: Hybrid seedling

Habit: Tall, growth vigorous.

Vegetative Characters: Stem young coppery; Leaf blade 7.0 x 4.3 cm, ovate, Yellow-green group 137 B, Fan-3; Thorn 1.0 cm long, straight.

Flower/Bract Characters: Flowering medium, borne all along the branches; Bracts 3.6 x 2.7 cm, ovate, Red-purple group 71 B, Fan-2, non-persistent; Star 0.7 cm dia., Yellow group 2 D, Fan-1 to Orange group 25 A, Fan-1.

45   **'Floribunda**

Species: *Bougainvillea glabra*

Parent: Seedling of *Bougainvillea glabra*

Habit: Tall, growth vigorous.

Vegetative Characters: Stem young coppery; Leaf blade 5.8 x 2.9 cm, elliptic, Green group 137 C, Fan-3; Thorn 0.8 cm long, straight.

Flower/Bract Characters: Flowering profuse, borne all along the branches; Bracts 4.0 x 2.9 cm, elliptic, Red-purple group 72 B, Fan-2, non-persistent; Star 0.7 cm dia., Green-yellow group 1 D, Fan-1.

46   **'Garnet Glory'**

Species: *Bougainvillea x buttiana*

Parent: Seedling of *Bougainvillea x buttiana*

Habit: Medium, growth vigorous.

Vegetative Characters: Stem young coppery; Leaf blade 8.0 x 5.0 cm, ovate, Green group 137 B, Fan-3; Thorn 1.2 cm long, straight.

Flower/Bract Characters: Flowering medium, at the end of the branches; Bracts 3.1 x 2.5 cm, ovate, Red-purple group 71 B, Fan-2, non-persistent; Star 0.7 cm dia., Green-yellow group 1 D, Fan-1.

47 **'Gillian Greensmith'**

Species: *Bougainvillea peruviana*

Parent: Hybrid Seedling

Habit: Drooping branches, growth vigorous.

Vegetative Characters: Stem young coppery; Leaf blade 5.2 x 3.1 cm, elliptic to ovate, Green group 137 C, Fan-3; Thorn 1.0 cm long, slightly curved.

Flower/Bract Characters: Flowering profuse, Bracts borne at the end and all along the branches; Bracts 3.6 x 2.5 cm, elliptic, Red-purple group 70 A, Fan-2, non-persistent; Star 0.6 cm dia., Red group 38 D, Fan-1.

48 **'Glabra'**

Species: *Bougainvillea glabra*

Parent: Seedling of *Bougainvillea glabra*

Habit: Dwarf, growth vigorous.

Vegetative Characters: Stem young green; Leaf blade 7.5 x 4.8 cm, elliptic, Yellow-green group 144 A, Fan-3; Thorn 1.3 cm long, curved.

Flower/Bract Characters: Flowering profuse, all along the branches; Bracts 4.1 x 3.0 cm, ovate, Red-purple group 62 C, Fan-2, persistent; Star 0.7 cm dia., Green-yellow group 1 A, Fan-1.

49 **'Glabra Magnifica'**

Species: *Bougainvillea glabra*

Parent: Seedling of *Bougainvillea glabra*

Habit: Medium, vigorous growth.

Vegetative Characters: Stem young greenish-coppery; Leaf blade 7.8 x 4.2 cm, elliptic, Green group 137 B, Fan-3; Thorn 1.3 cm long, curved.

Flower/Bract Characters: Flowering profuse, all along the branches; Bracts 3.0 x 1.6 cm, ovate, Red-purple group 72 B, Fan-2, persistent; Star 0.6 cm dia., Yellow group 3 D, Fan-1.

50 **'Glabra Magnifica Trilli'**

Species: *Bougainvillea glabra*

Parent: Seedling of *Bougainvillea glabra*

Habit: Medium, vigorous growth.

Vegetative Characters: Stem young greenish-coppery; Leaf blade 8.0 x 4.1 cm, elliptic, Green group 137 B, Fan-3; Thorn 1.2 cm long, curved.

Flower/Bract Characters: Flowering profuse, all along the branches; Bracts 4.1 x 3.0 cm, ovate, Red-purple group 72 B, Fan-2, persistent; Star 0.7 cm dia., Green-yellow group 1 D, Fan-1.

51   'Glabra Sanderiana'

Species: *Bougainvillea glabra*

Parent: Seedling of *Bougainvillea glabra*

Habit: Medium, vigorous growth.

Vegetative Characters: Stem young greenish-coppery; Leaf blade 7.5 x 4.1 cm, elliptic, Green group 137 A, Fan-3; Thorn 1.2 cm long, curved.

Flower/Bract Characters: Flowering profuse, all along the branches; Bracts 3.5 x 2.6 cm, ovate, Red-purple group 70 A, Fan-2, persistent; Star 0.7 cm dia., Green-yellow group 1 D, Fan-1.

52   'Gloriosus'

Species: *Bougainvillea peruviana*

Parent: Unknown

Habit: Tall, growth vigorous.

Vegetative Characters: Stem young greenish-coppery; Leaf blade 4.1 x 2.9 cm, ovate, Green group 137 B, Fan-3; Thorn 1.2 cm long, straight.

Flower/Bract Characters: Flowering profuse, at the end of the branches; Bracts 3.1 x 2.4 cm, ovate, Red-purple group 71 B, Fan-2, non-persistent; Star 0.6 cm dia., Green-yellow group 1 D, Fan-1.

53   'Gokul'

Species: *Bougainvillea x buttiana*

Parent: Unknown

Habit: Tall, growth vigorous.

Vegetative Characters: Stem young greenish-coppery; Leaf blade 11.1 x 6.7 cm, ovate, Green group 137 B, Fan-3; Thorn 1.2 cm long, straight.

Flower/Bract Characters: Flowering profuse, bracts borne all along the branches; Bracts 4.1 x 3.2 cm, ovate, Red-purple group 71 B, Fan-2, non-persistent; Star 0.7 cm dia., Green-yellow group 1 D, Fan-1.

54   'Golden Gaint'

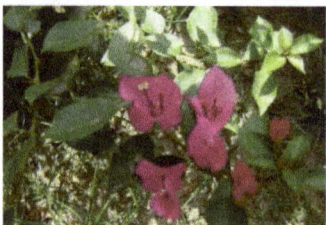

Species: *Bougainvillea peruviana*

Parent: Unknown

Habit: Medium, vigorous growth.

Vegetative Characters: Stem young coppery; Leaf blade 5.5 x 3.2 cm, ovate, Green group 137 B, Fan-3; Thorn 1.2 cm long, straight.

Flower/Bract Characters: Flowering profuse, at the end of the branches; Bracts 3.6 x 2.5 cm, ovate, Red-purple group 71 C, Fan-2, non-persistent; Star 0.7 cm dia., Green-yellow group 1 C, Fan-1.

55   'Golden Glory'

Species: *Bougainvillea x butttiana*

Parent: Unknown

Habit: Tall, growth vigorous.

Vegetative Characters: Stem young coppery; Leaf blade 3.1 x 1.8 cm, ovate, Green group 137 A, Fan-3; Thorn 1.3 cm long, straight.

Flower/Bract Characters: Flowering profuse, at the end of the branches; Bracts 3.2 x 2.7 cm, ovate, Red group 36 C, Fan-1, twisted, non-persistent; Star 0.6 cm dia., Yellow group 3 D, Fan-1.

56 **'Golden Glow'**

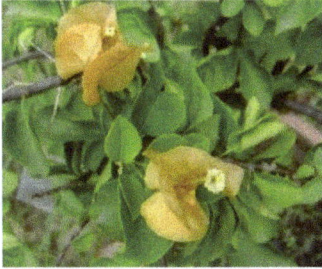

Species: *Bougainvillea x buttiana*

Parent: Bud sport of 'Mrs. Butt'

Habit: Tall, growth vigorous.

Vegetative Characters: Stem young greenish-coppery; Leaf blade 7.2 x 4.9 cm, ovate, Green group 137 C, Fan-3, curved; Thorn 1.2 cm long, curved.

Flower/Bract Characters: Flowering profuse, at the end of the branches; Bracts 2.9 x 2.5 cm, ovate, Yellow-orange group 16 C, Fan-1, non-persistent; Star 0.6 cm dia., Green-yellow group 1 D, Fan-1.

57 **'Gopal'**

Species: *Bougainvillea glabra*

Parent: Seedling of 'Formosa'

Habit: Medium, growth restricted.

Vegetative Characters: Stem young greenish-coppery; Leaf blade 5.9 x 4.8 cm, ovate, Green group 137 B, Fan-3; Thorn 1.1 cm long, straight.

Flower/Bract Characters: Flowering profuse, bracts borne all along the branches; Bracts 3.6 x 3.1 cm, ovate, Red-purple group 71 B, Fan-2, non-persistent; Star 0.6 cm dia., Green-yellow group 1 D, Fan-1.

58 **'Happiness'**

Species: *Bougainvillea spectabilis*

Parent: Seedling of *Bougainvillea spectabilis*

Habit: Medium, Growth vigorous.

Vegetative Characters: Stem young greenish-coppery; Leaf blade 6.6 x 4.4 cm, ovate, Green group 137 B, Fan-3; Thorn 1.3 cm long, slightly curved.

Flower/Bract Characters: Flowering profuse, Bracts at the end of the branches; Bracts 3.1 x 2.3 cm, ovate, Red-purple group 60 B, Fan-2, non-persistent; Star 0.7 cm dia., Green-yellow group 1 D, Fan-1.

59 **'Hawaiian Beauty'**

Species: *Bougainvillea x buttiana*

Parent: Bud sport of 'Hawaiian White'

Habit: Medium, growth vigorous.

Vegetative Characters: Stem young greenish-coppery; Leaf blade 5.1 x 3.5 cm, ovate, Green group 137 B, Fan-3, variegated (creamy yellow margin); Thorn 1.2 cm long, straight.

Flower/Bract Characters: Flowering medium, at the end of the branches; Bracts 2.7 x 2.3 cm, ovate, Green-white group 157 C, Fan 4, non-persistent; Star 0.6 cm dia., Green-yellow group 1 C, Fan-1.

60 'Hawaiian White'

Species: *Bougainvillea x buttiana*

Parent: Unknown

Habit: Medium, growth moderately.

Vegetative Characters: Stem young greenish-coppery; Leaf blade 5.6 x 3.8 cm, ovate, Green group 137 B, Fan-3; Thorn 1.2 cm long, straight.

Flower/Bract Characters: Flowering medium, at the end of the branches; Bracts 2.7 x 2.3 cm, ovate, Green-white group 157 C, Fan 4, twisted, non-persistent; Star 0.6 cm dia., Green-yellow group 1 C, Fan-1.

61 'Himani' / 'Miss. Alice'

Species: *Bougainvillea* sp.

Parent: Unknown

Habit: Medium, growth vigorous.

Vegetative Characters: Stem young green; Leaf blade 12.5 x 5.0 cm, elliptic, Yellow-green group 144 A, Fan-3; Thorn mainly absent but when present 0.6 cm long, curved.

Flower/Bract Characters: Flowering profuse, bracts borne all along the branches; Bracts 6.3 x 3.2 cm, ovate, Green White group 137 B, Fan 4, non-persistent; Star 0.8 cm dia., Green-yellow group 1 C, Fan-1.

62 'Isabel Greensmith'

Species: *Bougainvillea peruviana*

Parent: Hybrid Seedling

Habit: Tall, growth vigorous.

Vegetative Characters: Stem young greenish coppery; Leaf blade 5.8 x 3.2 cm, elliptic to ovate, Green group 137 B, Fan-3; Thorn 1.8 cm long, straight.

Flower/Bract Characters: Flowering profuse, at the end and all along the branches; Bracts 3.3 x 1.8 cm, ovate to elliptic, Orange-red group 34 A, Fan-1 to Red-purple group 70 A, Fan-2, non-persistent; Star 0.6 cm dia., change Yellow group 3 D, Fan-1 to Yellow group 13 C, Fan-1.

63 'Jaya'

Species: *Bougainvillea peruviana*

Parent: Unknown

Habit: Medium, growth moderately.

Vegetative Characters: Stem young greenish coppery; Leaf blade 6.1 x 4.9 cm, ovate, Green group 137 B, Fan-3; Thorn 1.3 cm long, straight.

Flower/Bract Characters: Flowering profuse, bracts borne all along the branches; Bracts 3.2 x 2.7 cm, ovate, Red-yellow group 39 B, Fan-1, non-persistent; Star 0.6 cm dia., Yellow group 3 D, Fan-1.

64 'Jayalakshmi'

Species: *Bougainvillea x buttiana*

Parent: Seedling of 'Meera'

Habit: Tall, growth vigorous.

Vegetative Characters: Stem young greenish coppery; Leaf blade 5.4 x 4.5 cm, ovate, Green group 137 B, Fan-3; Thorn 1.0 cm long, straight.

Flower/Bract Characters: Flowering profuse, 1/3 of the branches; Bracts 3.1 x 2.2 cm, ovate, Red-purple group 71 C, Fan-2, non-persistent; Star 0.5 cm dia., Green-yellow group 1 D, Fan-1.

65 **'Joe de Lovera'**

Species: *Bougainvillea glabra*

Parent: Unknown

Habit: Medium, growth vigorous.

Vegetative Characters: Stem young coppery; Leaf blade 7.2 x 4.5 cm, ovate, Green group 137AB, Fan-3; Thorn 1.0 cm long, slightly curved.

Flower/Bract Characters: Flowering profuse, at the end of the branches; Bracts 3.0 x 2.2 cm, ovate, Red-purple group 64 B, Fan-2, non-persistent; Star 0.6 cm dia., Yellow group 2 D, Fan-1.

66 **'Jubilee'**

Species: *Bougainvillea spectabilis*

Parent: Unknown

Habit: Tall, growth vigorous.

Vegetative Characters: Stem young coppery; Leaf blade 7.7 x 5.5 cm, broadly ovate, Green group 137 B, Fan-3; Thorn 1.2 cm long, slightly curved.

Flower/Bract Characters: Flowering sparse, at the end of the branches; Bracts 3.3 x 2.7 cm, ovate, Red-purple group 68 A, Fan-2, base of bracts is white in colour, non-persistent; Star non-prominent.

67 **'Killie Campbell'**

Species: *Bougainvillea x buttiana*

Parent: Seedling of *Bougainvillea x buttiana*

Habit: Dwarf, drooping, growth vigorous.

Vegetative Characters: Stem young greenish brown; Leaf blade 3.6 x 2 cm, ovate, Green group 137 B, Fan-3; Thorn 1.7 cm long, straight.

Flower/Bract Characters: Flowering profuse, Bracts borne all along the branches; Bracts 3.5 x 1.8 cm, elliptic, Red-purple group 71 B, Fan-2, non-persistent; Star 0.7 cm dia., Green-yellow group 1 D, Fan-1.

68 **'Krumbiegel'**

Species: *Bougainvillea peruviana*

Parent: Seedling of 'Mahatma Gandhi'

Habit: Tall, drooping, growth vigorous.

Vegetative Characters: Stem young greenish brown; Leaf blade 6.2 x 4.5 cm, ovate, Green group 137 C, Fan-3; Thorn 1.4 cm long, straight.

Flower/Bract Characters: Flowering profuse, 1/3 of the branches; Bracts 4.1 x 3.2 cm, ovate, Red-purple group 72 B, Fan-2, twisted, non-persistent; Star 0.5 cm dia., Yellow group 3 D, Fan-1.

69 **'Lady Hope'**

Species: *Bougainvillea peruviana*

Parent: Hybrid seedling of *Bougainvillea peruviana*

Habit: Tall, growth vigorous.

Vegetative Characters: Stem young greenish coppery; Leaf blade 7.1 x 4.2 cm, elliptic, Green group 137 B, Fan-3; Thorn 1.1 cm long, slightly curved.

Flower/Bract Characters: Flowering profuse, borne all along the branches; Bracts 4.1 x 2.8 cm, ovate, Red-purple group 70 B, Fan-2, non-persistent; Star 0.7 cm dia., Green-yellow group 1 D, Fan-1.

70 'Lady Hudson of Ceylon'

Species: *Bougainvillea x buttiana*

Parent: Seedling of *Bougainvillea x buttiana*

Habit: Tall, growth intermediate.

Vegetative Characters: Stem young greenish coppery; Leaf blade 6.2 x 4.1 cm, ovate, Yellow-green group 144 A, Fan-3; Thorn 1.1 cm long, curved.

Flower/Bract Characters: Flowering profuse, at the end of the branches; Bracts 3.7 x 2.5 cm, ovate, Red-purple group 62 C, Fan-2, non-persistent; Star 0.5 cm dia., Yellow group 7 D, Fan-1.

71 'Lady Mary Baring'

Species: *Bougainvillea x buttiana*

Parent: Bud Sport of 'Golden Glow'

Habit: Tall, vigorous growth.

Vegetative Characters: Stem young green; Leaf blade 6.7 x 4.7 cm, broadly ovate, Yellow-green group 144 A, Fan-3; Thorn 1.4 cm long, slightly curved.

Flower/Bract Characters: Flowering profuse, at the end of the branches; Bracts 2.8 x 2.1 cm, ovate, Yellow-orange group 20 A, Fan-1, non-persistent; Star 0.6 cm dia., Yellow group 3 D, Fan-1.

72 'Lady Mountbatten'

Species: *Bougainvillea spectabilis*

Parent: Hybrid seedling of *B. spectabilis*

Habit: Tall, vigorous growth.

Vegetative Characters: Stem young greenish coppery; Leaf blade 5.9 x 4.1 cm, ovate, Green group 137 C, Fan-3; Thorn 1.6 cm long, slightly curved.

Flower/Bract Characters: Flowering profuse, borne all along the branches; Bracts 4.1 x 2.9 cm, ovate, Red-purple group 72 B, Fan-2, non-persistent; Star 0.6 cm dia., Yellow group 4 C, Fan-1.

73 'Lady Richard'

Species: *Bougainvillea glabra*

Parent: Seedling of *Bougainvillea glabra*

Habit: Tall, vigorous growth.

Vegetative Characters: Stem young greenish coppery; Leaf blade 5.9 x 3.1 cm, ovate, Green group 137 C, Fan-3; thorn 1.4 cm long, straight.

Flower/Bract Characters: Flowering intermediate, 1/2 of the branches, Bracts small 3.8 x 2.5 cm, ovate, Red-purple group 59 C, Fan-2, non-persistent; Star 0.6 cm dia., Yellow group 3 C, Fan-1.

74 'Lateritia'

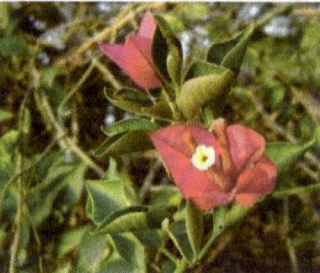

Species: *Bougainvillea spectabilis*

Parent: Hybrid seedling of *B. spectabilis*

Habit: Medium, growth intermediate.

Vegetative Characters: Stem young greenish coppery; Leaf blade 5.2 x 4.2 cm, ovate or broadly ovate, Green group 137 A, Fan-3; Thorn 1.4 cm long, slightly curved.

Flower/Bract Characters: Flowering sparse, borne all along the branches; Bracts 2.9 x 1.8 cm, ovate, Red-purple group 63 A, Fan-2, non-persistent; Star 0.6 cm dia., Green-yellow group 1 C, Fan-1.

75 **'Lilac Perfection'**

Species: *Bougainvillea glabra*

Parent: Unknown

Habit: Dwarf, growth restricted.

Vegetative Characters: Stem young green; Leaf blade 8.5 x 4.0 cm, elliptic, Green group 137 C, Fan-3; Thorn 1.1 cm long, straight.

Flower/Bract Characters: Flowering profuse, Bracts borne all along the branches; Bracts 3.5 x 1.5 cm, elliptic, Red-purple group 62 C, Fan-2, non-persistent; Star 0.7 cm dia., Green-yellow group 1 C, Fan-1.

76 **'Lilac Puff'**

Species: *Bougainvillea glabra*

Parent: Unknown

Habit: Dwarf, growth restricted.

Vegetative Characters: Stem young green; Leaf blade 6.2 x 3.0 cm, elliptic, Green group 137 B, Fan-3; Thorn 0.7 cm long, straight.

Flower/Bract Characters: Flowering medium, Bracts borne at the tip of the branches; Bracts 2.5 x 1.5 cm, narrowly or medium ovate, Purple group 75 A, Fan-2, persistent; Star 0.5 cm dia., Green-yellow group 1 C, Fan-1.

77 **'Limousine'**

Species: *Bougainvillea peruviana*

Parent: Unknown

Habit: Dwarf, growth restricted.

Vegetative Characters: Stem young greenish coppery; Leaf blade 4.5 x 2.9 cm, ovate, Yellow-green group 144 A, Fan-3; Thorn 1.0 cm long, straight.

Flower/Bract Characters: Flowering profuse, borne all along the branches; Bracts 3.3 x 2.8 cm, elliptic, Red-purple group 60 C, Fan-2, non-persistent; Star 0.6 cm dia., Green-yellow group 1 C, Fan-1.

78 **'Little Node'**

Species: *Bougainvillea peruviana*

Parent: Unknown

Habit: Medium, growth moderate

Vegetative Characters: Stem young green; Leaf blade 8.2 x 4.4 cm, ovate, Green group 137 C, Fan-3; Thorn 1.2 cm long, curved.

Flower/Bract Characters: Flowering profuse, borne at the end of branches; Bracts 3.2 x 2.1 cm, ovate, Red-purple group 71 B, Fan-2, non-persistent; Star 0.6 cm dia., Yellow group 4 D, Fan-1.

79 **'Los Banos Beauty'**

Species: *Bougainvillea x buttiana*

Parent: Bud sport of 'Pink Beauty'

Habit: Tall, growth vigorous.

Vegetative Characters: Stem young greenish coppery; Leaf blade 5.6 x 4.5 cm, ovate, Green group 137 C, Fan-3; Thorn 0.8 cm long, slightly curved.

Flower/Bract Characters: Flowering profuse, at the end of the branches; Bracts 2.8 x 1.9 cm, elliptic to ovate, Red-purple group 68 B, Fan-2, persistent; Floral tube absent.

80 'Los BanosVariegata'

Species: *Bougainvillea x buttiana*

Parent: Mutant of ' Los Banos Beauty'

Habit: Intermediate, growth moderate.

Vegetative Characters: Stem young coppery; Leaves variegated (creamy white at margin), leaf blade 5.9 x 4.6 cm, broadly ovate, Green group 138 A, Fan-3; Thorn 0.7 cm long, straight.

Flower/Bract Characters: Flowering profuse, at the end of the branches; Bracts 3.0 x 1.9 cm, elliptic to ovate, Red-purple group 70 C, Fan-2, persistent; Floral tube absent.

81 'Los BanosVariegata 'Jayanti'

Species: *Bougainvillea x buttiana*

Parent: Mutant of ' Los Banos Beauty'

Habit: Growth retarded.

Vegetative Characters: Stem young greenish coppery; Leaf blade 5.6 x 4.5 cm, broadly ovate, Green group 137 C, Fan-3, Mutant showed two types of variegated leaves. One type is mosaic which has four visible colours i.e. Olive green (Yellow-green Group 146 B, Fan-3), Pale yellow (Yellow-green Group 146 D, Fan-3), Deep Green (Yellow-green Group 146 A, Fan-3) and Light Yellow (Yellow-green Group 154 D, Fan-3), mixed with different shades of green. The second type of leaf has green centre with creamish white margin. Foliage of this type has three major colour combinations i.e. Light Yellow (Yellow-Orange Group 19 C, Fan-1), Green (Green Group 147 B, Fan-3) and Light Green (Yellow- Green Group 148 C, Fan-3); Thorn 0.9 cm long, slightly curved.

Flower/Bract Characters: Flowering profuse, at the end of the branches, bract 3.0 x 1.9 cm, elliptic to ovate, Red-purple group 70 C, Fan-2, persistent; Floral tube absent.

82 Los Banos Variegata 'Silver Margin'

Species: *Bougainvillea x buttiana*

Parent: Mutant of 'Los Banos Beauty'

Habit: Dwarf, drooping, growth intermediate.

Vegetative Characters: Stem young coppery; Leaf blade 7.6 x 5.1 cm, ovate, Green group 137 C, Fan-3, leaves variegated (light green speckled at margins); Thorn 0.8 cm long, curved.

Flower/Bract Characters: Flowering profuse, at the end of the branches; Bracts 2.5 x 1.6 cm, elliptic to ovate, Red-purple group 68 B, Fan-2, persistent; Floral tube absent.

83 'Louise Wathen'

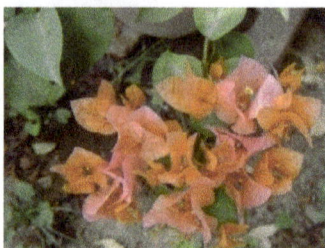

Species: *Bougainvillea x buttiana*

Parent: Bud sport of 'Mrs. Butt'

Habit: Tall, growth vigorous.

Vegetative Characters: Stem young greenish brown; Leaf blade 5.5 x 4.2 cm, ovate, Green group 137 C, Fan-3; Thorn 1.0 cm long, straight.

Flower/Bract Characters: Flowering medium, at the end of the branches; Bracts 3.3 x 2.8 cm, ovate, Red group 37 A, Fan-1, non-persistent; Star 0.5 cm dia., Yellow Orange group 14 D, Fan-1.

84 **'Louise Wathen Variegata'**

Species: *Bougainvillea x buttiana*

Parent: Bud sport 'Louise Wathen'

Habit: Tall, growth vigorous.

Vegetative Characters: Stem young greenish coppery; Leaf blade 6.5 x 5.4 cm, broadly ovate, Green group 138 B, Fan-3, leaves variegated (white at margin); Thorn 1.0 cm long, straight.

Flower/Bract Characters: Flowering medium, at the end of the branches; Bracts 3.5 x 2.8 cm, ovate, Red group 62 A, Fan-2, non-persistent; Star 0.7 cm dia., Yellow group 2 D, Fan-1.

85 **'Lucifer Red'**

Species: *Bougainvillea glabra*

Parent: Unknown

Habit: Medium, growth vigorous.

Vegetative Characters: Stem young coppery; Leaf blade 6.9 x 4.6 cm, ovate, Green group 137 C, Fan-3; Thorn 1.0 cm long, straight to slightly curved.

Flower/Bract Characters: Flowering profuse, bracts borne all along the branches; Bracts 2.9 x 2.4 cm, ovate, Red-purple group 71 B, Fan-2, non-persistent; Star 0.7 cm dia., Green-yellow group 1 C, Fan-1.

86 **'Mackakos'**

Species: *Bougainvillea peruviana*

Parent: Unknown

Habit: Tall, growth vigorous.

Vegetative Characters: Stem young greenish coppery; Leaf blade 5.9 x 3.0 cm, ovate, Green group 137 A, Fan-3; Thorn 1.2 cm long, straight.

Flower/Bract Characters: Flowering profuse, borne all along the branches; Bracts 3.4 x 2.5 cm, ovate, Red-purple group 71 C, Fan-2, non-persistent; Star 0.7 cm dia., Green-yellow group 1 D, Fan-1.

87 **'Mahara'**

Species: *Bougainvillea x buttiana*

Parent: Bud sport of 'Mrs. Butt'

Habit: Tall, growth vigorous.

Vegetative Characters: Stem young greenish coppery; Leaf blade 6.1 x 4.2 cm, ovate, Green group 137 C, Fan-3; Thorn 1.3 cm long, curved.

Flower/Bract Characters: Flowering profuse, at the end of the branches; Bracts 2.0 x 1.2 cm, elliptic to ovate, Red-purple group 70 A, Fan-2, persistent; Floral tube absent.

88 **Mahara 'Abnormal Leaves'**

Species: *Bougainvillea x buttiana*

Parent: Bud sport of 'Mahara'

Habit: Tall, growth vigorous.

Vegetative Characters: Stem young green; Leaf blade 8.3 x 6.2 cm, ovate, abnormal in shape (margin is unevenly undulated and leaf lamina is asymmetrical in shape), variegated (light green all over the surface of leaves), Green group 137 C, Fan-3; Thorn 1.4 cm long, curved.

Flower/Bract Characters: Flowering profuse, at the end of the branches; Bracts 2.2 x 1.2 cm, elliptic to ovate, Red-purple group 71 D, Fan-2, non-persistent; Floral tube absent.

89 'Mahatma Gandhi'

Species: *Bougainvillea peruviana*

Parent: Seedling of 'Princess Margaret Rose'

Habit: Tall, growth vigorous.

Vegetative Characters: Stem young greenish coppery; Leaf blade 6.9 x 4.1 cm, ovate, Green group 137 A, Fan-3; Thorn 1.0 cm long, straight.

Flower/Bract Characters: Flowering profuse, all along the branches; Bracts 3.9 x 2.7 cm, elliptic, Red-purple group 72 B, Fan-2, non-persistent; Star 0.7 cm dia., Green-yellow group 1 D, Fan-1.

90 'Manohar Chandra'

Species: *Bougainvillea x buttiana*

Parent: Hybrid seedling of *Bougainvillea x buttiana*

Habit: Medium, growth restricted.

Vegetative Characters: Stem young green; Leaf blade 6.9 x 4.6 cm, ovate to elliptic, Green group 137 A, Fan-3; Thorn 1.1 cm long, slightly curved.

Flower/Bract Characters: Flowering sparse, at the end of the branches; Bracts 3.4 x 2.3 cm, ovate, Red-purple group 70 A, Fan-2, non-persistent; Star 0.6 cm dia., Yellow group 4 D, Fan-1.

91 'Manohar Chandra Variegata'

Species: *Bougainvillea x buttiana*

Parent: Bud sport of 'Manohar Chandra'

Habit: Medium, growth restricted.

Vegetative Characters: Stem young green; Leaf blade 6.9 x 4.6 cm, ovate to elliptic, Green group 137 A, Fan-3; Thorn 1.1 cm long, slightly curved.

Flower/Bract Characters: Flowering sparse, at the end of the branches; Bracts 3.4 x 2.3 cm, ovate, Red-purple group 70 A, Fan-2, non-persistent; Star 0.6 cm dia., Yellow group 4 D, Fan-1.

92 'Margery Lloyd'

Species: *Bougainvillea spectabilis*

Parent: Hybrid Seedling of *Bougainvillea spectabilis*

Habit: Tall, growth vigorous.

Vegetative Characters: Stem young green; Leaf blade 7.9 x 4.7 cm, ovate, Yellow-green group 144 A, Fan-3, hairy; Thorn 1.9 cm long, curved.

Flower/Bract Characters: Flowering profuse, all along the branches; Bracts 4.4 x 2.3 cm, elliptic, Red group 37 A, Fan-1, non-persistent; Star 0.6 cm dia., Green-yellow group 1 D, Fan-1.

93 'Mary Palmer'

Species: *Bougainvillea peruviana*

Parent: Bud sport of 'Mrs. H. C. Buck'

Habit: Tall, growth vigorous.

Vegetative Characters: Stem young greenish coppery; Leaf blade 8.0 x 5.9 cm, broadly ovate, Yellow-green group 144 A, Fan-3; Thorn 1.4 cm long, slightly curved.

Flower/Bract Characters: Flowering profuse, all along the branches; Bracts 3.6 x 2.9 cm, ovate, double colour, Red-purple group 72 B, Fan-2, White group 157 C, Fan 4, non-persistent; Star 0.6 cm dia., Yellow group 3 D, Fan-1.

94 **'Mary Palmer Special'**

Species: *Bougainvillea peruviana*

Parent: Hybrid seedling 'Dr. B.P. Pal' X 'Princess Margaret Rose'

Habit: Tall, growth vigorous.

Vegetative Characters: Stem young greenish coppery; Leaf blade 8.2 x 6.1 cm, broadly ovate, Green group 138 A, Fan-3; Thorn 1.7cm in long, slightly curved.

Flower/Bract Characters: Flowering profuse, all along the branches, bract 4.5 x 3.7 cm, three type of bract colour - Red-purple group 70 A, Fan-2, Green-white group 157 C, Fan 4, Green-white with Red-purple, non-persistent; Star 0.8 in dia., Green-yellow group 1 D, Fan-1.

95 **'Mrs. Bakery'**

Species: *Bougainvillea x buttiana*

Parent: Hybrid seedling of *B. peruviana x B. glabra*

Habit: Tall, growth vigorous.

Vegetative Characters: Stem young greenish coppery; Leaf blade 4.5 x 3.5 cm, Medium ovate, Green group 137 C, Fan-3; Thorn 1.0 cm long, slightly curved.

Flower/Bract Characters: Flowering profuse, at the end of the branches; Bracts 3.5 x 3.1 cm, ovate, Red-purple group 64 B, Fan-2, non-persistent; Star 0.6 cm dia., Green-yellow group 1 D, Fan-1.

96 **'Mrs. Butt'**

Species: *Bougainvillea x buttiana*

Parent: Hybrid seedling of B. peruvianax B. glabra

Habit: Tall, growth vigorous.

Vegetative Characters: Stem young greenish coppery; Leaf blade 5.9 x 4.7 cm, broadly ovate, Green group 137 B, Fan-3; Thorn 1.2 cm long, straight.

Flower/Bract Characters: Flowering profuse, at the end of the branches; Bracts 3.6 x 3.0 cm, ovate, Red-purple group 60 C, Fan-2, non-persistent; Star 0.7 cm dia., Green-yellow group 1 D, Fan-1.

97 **'Mrs. Delux Perry'**

Species: *Bougainvillea glabra*

Parent: Unknown

Habit: Tall, growth vigorous

Vegetative Characters: Stem young coppery; Leaf blade 8.2 x 5.7 cm, ovate, Green group 137 A, Fan-3; Thorn 1.4 cm long, straight to slightly curved.

Flower/Bract Characters: Flowering profuse, at the end of the branches; Bracts 3.3 x 2.4 cm, ovate, Red-purple group 67 A, Fan-2, non-persistent; Star 0.6 cm dia., Yellow group 4 D, Fan-1.

98 **'Mrs. Eva Variegata' (Light Purple)**

Species: *Bougainvillea glabra*

Parent: Unknown

Habit: Tall, growth vigorous

Vegetative Characters: Stem young green; Leaf blade 7.5 x 5.7 cm, ovate, variegated, Green group 138 A, Fan-3, Yellow group 2 D, Fan-1; Thorn 1.9 cm long, curved.

Flower/Bract Characters: Flowering profuse, 1/3rd of the branches; Bracts 3.8 x 2.4 cm, elliptic, Purple group 75 A, Fan-2, non-persistent; Star 0.9 cm dia., Green-yellow group 1 C, Fan-1.

99 **'Mrs. Fraser'**

Species: *Bougainvillea spectabilis*

Parent: Unknown

Habit: Tall, growth vigorous.

Vegetative Characters: Stem young greenish coppery; Leaf blade 6.9 x 4.6 cm, ovate, Green group 137 C, Fan-3; Thorn 0.9 cm long, straight to slightly curved.

Flower/Bract Characters: Flowering profuse, 1/3rd of the branches; Bracts 3.6 x 2.2 cm, ovate, Red group 45 D, Fan-1, non-persistent; Star 0.7 cm dia., Green-yellow group 1 C, Fan-1.

100 **'Mrs. H. C. Buck'**

Species: *Bougainvillea peruviana*

Parent: Hybrid seedling of 'Princess Margaret Rose'

Habit: Tall, growth vigorous.

Vegetative Characters: Stem young greenish coppery; Leaf blade 8.6 x 5.3 cm, ovate, Green group 137 B, Fan-3; Thorn 1.1 cm long, straight.

Flower/Bract Characters: Flowering profuse, all along the branches; Bracts 4.2 x 2.8 cm, ovate, Red-purple group 70 B, Fan-2, non-persistent; Star 0.7 cm dia., Green-yellow group 1 D, Fan-1.

101 **'Mrs. McClean'**

Species: *Bougainvillea x buttiana*

Parent: Bud sport of 'Mrs. Butt'

Habit: Tall, growth vigorous.

Vegetative Characters: Stem young greenish coppery; Leaf blade 4.9 x 3.6 cm, ovate, Green group 137 C, Fan-3; Thorn 1.5 cm long, slightly curved.

Flower/Bract Characters: Flowering profuse, at the end of the branches; Bracts 3.0 x 2.4 cm, ovate, Red group 55 B, Fan-1, non-persistent; Star 0.6 cm dia., Green-yellow group 1 D, Fan-1.

102 **'Mrs. Oliver Perry'**

Species: *Bougainvillea x buttiana*

Parent: Hybrid seedling

Habit: Tall, growth vigorous.

Vegetative Characters: Stem young coppery; Leaf blade 7.9 x 4.7 cm, ovate, Green group 137 A, Fan-3; Thorn 1.2 cm long, straight.

Flower/Bract Characters: Flowering medium, at the end of the branches; Bracts 3.6 x 2.2 cm, elliptic, Red-purple group 70 A, Fan-2, non-persistent; Star 0.7 cm dia., Yellow group 2 D, Fan-1 and Red-purple group 62 C, Fan-2.

103 **'Mrs. R. B. Carrick'**

Species: *Bougainvillea spectabilis*

Parent: Seedling of *B. spectabilis*

Habit: Tall, growth vigorous.

Vegetative Characters: Stem young coppery; Leaf blade 7.94 x 4.8 cm, ovate, Green group 137 A, Fan-3; Thorn 1.1 cm long, straight.

Flower/Bract Characters: Flowering profuse, all along the branches; Bracts 4.2 x 2.9 cm, ovate, Red-purple group 72 B, Fan-2, non-persistent; Star 0.7 cm dia., Green-yellow group 1 D, Fan-1.

104 **'New Red'**

Species: *Bougainvillea peruviana*

Parent: Unknown

Habit: Tall, growth vigorous.

Vegetative Characters: Stem young green; Leaf blade 7.1 x 4.4 cm, elliptic, Green group 137 B, Fan-3; Thorn 1.0 cm long, straight.

Flower/Bract Characters: Flowering profuse, 1/2 along the branches; Bracts 2.9 x 2.2 cm, ovate, Red-purple group 64 B, Fan-2, non-persistent; Star 1.1 cm dia., Green-yellow group 1 C, Fan-1.

105 **'Nigrette'**

Species: *Bougainvillea glabra*

Parent: Unknown

Habit: Medium, vigorous growth.

Vegetative Characters: Stem young green; Leaf blade 7.2 x 3.7 cm, elliptic, Green group 137 B, Fan-3; Thorn 1.3cm long, curved.

Flower/Bract Characters: Flowering profuse, all along the branches; Bracts 3.4 x 2.3 cm, ovate, Red-purple group 70 A, Fan-2, non-persistent; Star 0.6 cm dia., Green-yellow group 1 D, Fan-1.

106 **'Nirmal'**

Species: *Bougainvillea x buttiana*

Parent: Bud sport of 'Mrs. McClean'

Habit: Medium, growth restricted.

Vegetative Characters: Stem young green; Leaf blade 7.0 x 5.6 cm, ovate, Green group 137 B, Fan-3; Thorn 0.8 cm long, curved.

Flower/Bract Characters: Flowering medium, at the end of the branches; Bracts 3.5 x 2.8 cm, ovate, Red group 32 D, Fan-1, non-persistent; Star 0.7 cm dia., Green-yellow group 1 D, Fan-1.

107 **'Odisee'**

Species: *Bougainvillea peruviana*

Parent: Bud sport of 'Mary Palmer'

Habit: Medium, growth vigorous.

Vegetative Characters: Stem young green; Leaf blade 9.8 x 6.8 cm, ovate, Yellow-green group 137 B, Fan-3; Thorn 1.1 cm long, curved.

Flower/Bract Characters: Flowering profuse, 1/2 along the branches; Bracts 4.0 x 3.3 cm, ovate, Purple group 75 B, Fan-2, non-persistent; Star 0.7 cm dia., Yellow group 3 D, Fan-1.

108 **'Padmi'**

Species: *Bougainvillea x buttiana*

Parent: Seedling of 'Brasiliensis'

Habit: Medium, growth moderately vigorous.

Vegetative Characters: Stem young coppery; Leaf blade 6.1 x 4.0 cm, ovate, Green group 137 B, Fan-3; Thorn 1.9 cm long, slightly curved.

Flower/Bract Characters: Flowering profuse, at the end of the branches; Bracts 3.1 x 2.5 cm, ovate, Red-purple group 70 B, Fan-2, non-persistent; Star 0.6 cm dia., Yellow group 3 D, Fan-1.

109 'Palekar'

Species: *Bougainvillea peruviana*

Parent: Seedling of *Bougainvillea peruviana*

Habit: Tall, drooping, growth vigorous.

Vegetative Characters: Stem young greenish coppery; Leaf blade 11.1 x 6.0 cm, ovate, Green group 137 A, Fan-3; Thorn 1.2 cm long, slightly curved.

Flower/Bract Characters: Flowering profuse, all along the branches; Bracts 4.0 x 2.6 cm, ovate, Red-purple group 64 B, Fan-2, non-persistent; Star 0.7 cm dia., Green-yellow group 1 C, Fan-1.

110 'Pallavi'

Species: *Bougainvillea x buttiana*

Parent: Mutant of 'Roseville's Delight'

Habit: Intermediate, plant growth vigorous.

Vegetative Characters: Stem young green; Leaf blade 9.1 x 4.0 cm, ovate, variegated (yellowish green), Yellow-green group 144 B, Fan-3 and Green group 138 B, Fan-3; Thorn 1.0 cm long, straight.

Flower/Bract Characters: Flowering profuse, at the end of the branches; Bracts 2.6 x 1.2 cm, elliptic, Red-purple group 62 A, Fan-2, persistent; Star absent.

111 'Partha'

Species: *Bougainvillea peruviana*

Parent: Seedling of *Bougainvillea peruviana*

Habit: Tall, drooping, growth vigorous.

Vegetative Characters: Stem young green; Leaf blade 7.5 x 4.6 cm, ovate to elliptic, Yellow-green group 144 A, Fan-3; Thorn 1.5 cm long, curved.

Flower/Bract Characters: Flowering profuse, all along the branches; Bracts 4.2 x 2.7 cm, ovate, Red-purple group 72 B, Fan-2, non-persistent; Star 0.8 cm dia., Green-yellow group 1 C, Fan-1.

112 'Parthasarthy'

Species: *Bougainvillea peruviana*

Parent: Bud sport of 'Partha'

Habit: Medium, drooping, growth vigorous.

Vegetative Characters: Stem young greenish coppery; Leaf blade 6.3 x 3.8 cm, elliptic, variegated (cream yellow patches), Green group 137 C, Fan-3; Thorn 1.3 cm long, curved.

Flower/Bract Characters: Flowering sparse, at the end of the branches; Bracts 4.1 x 2.9 cm, ovate, Red-purple group 70 B, Fan-2, non-persistent; Star 0.8 cm dia., Yellow group 2 D, Fan-1.

113 'Philips'

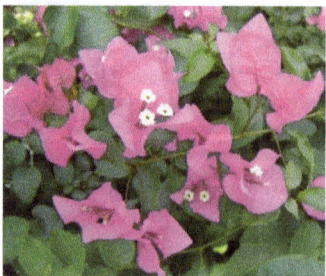

Species: *Bougainvillea spectabilis*

Parent: Unknown

Habit: Tall, growth vigorous.

Vegetative Characters: Stem young coppery; Leaf blade 8.2 x 4.9 cm, ovate, Green group 137 C, Fan-3; Thorn 1.5 cm long, slightly curved.

Flower/Bract Characters: Flowering profuse, all along the branches; Bracts 4.5 x 3.6 cm, ovate, Red-purple group 70 B, Fan-2, non-persistent; Star 0.9 cm dia., Green-yellow group 1 D, Fan-1.

114　**'Pisil'**

Species: *Bougainvillea peruviana*

Parent: Unknown

Habit: Medium growth restricted.

Vegetative Characters: Stem young greenish coppery; Leaf blade 6.7 x 4.4 cm, ovate, Green group 137 B, Fan-3; Thorn 1.2 cm long, straight.

Flower/Bract Characters: Flowering profuse, at the end of the branches; Bracts 3.3 x 2.5 cm, ovate, change from Red group 53 C, Fan-1 to Red-purple group 61 B, Fan-2, non-persistent; Star 0.6 cm dia., Green-yellow group 1 C, Fan-1.

115　**'Pixie'**

Species: *Bougainvillea x buttiana*

Parent: Unknown

Habit: Dwarf, growth moderately.

Vegetative Characters: Stem young greenish coppery; Leaf blade 4.9 x 3.0 cm, ovate, Green group 137 C, Fan-3; Thorn 1.4 cm long, slightly curved.

Flower/Bract Characters: Flowering profuse, at middle of the branches (auxiliary); Bracts 2.2 x 1.5 cm, ovate, Red-purple group 70 B, Fan-2, non-persistent; Star 0.6 cm dia., Yellow group 2 D, Fan-1.

116　**'Pixie Orange'**

Species: *Bougainvillea x buttiana*

Parent: Unknown

Habit: Dwarf, growth moderately.

Vegetative Characters: Stem young greenish coppery; Leaf blade 4.5 x 2.5 cm, ovate, Green group 137 B, Fan-3; Thorn 1.2 cm long, slightly curved.

Flower/Bract Characters: Flowering profuse, at middle of the branches (auxiliary); Bracts 2.3 x 1.4 cm, ovate, Red group 48 D, Fan-1, non-persistent; Star 0.6 cm dia., Green-yellow group 1 D, Fan-1.

117　**'Pixie Pink'**

Species: *Bougainvillea x buttiana*

Parent: Unknown

Habit: Dwarf, growth moderately.

Vegetative Characters: Stem young green; Leaf blade 4.9 x 3.0 cm, ovate, Green group 137 C, Fan-3; Thorn 1.2 cm long, slightly curved.

Flower/Bract Characters: Flowering sparse, at middle of the branches (auxiliary); Bracts 2.8 x 1.9 cm, ovate, Red-purple group 73 C, Fan-2, non-persistent; Star 0.5 cm dia., Yellow group 2 D, Fan-1.

118　**'Pixie Queen'**

Species: *Bougainvillea x buttiana*

Parent: Mutant of 'Pixie Pink'

Habit: Dwarf, growth moderately.

Vegetative Characters: Stem young green; Leaf blade 4.5 x 2.8 cm, ovate, variegated, speckled pattern, Yellow-green group 144 A, Fan-3, Yellow group 9 D, Fan-1; Thorn 1.1 cm long, straight to slightly curved.

Flower/Bract Characters: Flowering moderately, at middle of the branches (auxiliary); Bracts 2.4 x 1.4 cm, ovate, Red-purple group 73 C, Fan-2, non-persistent; Star 0.5 cm dia., Yellow group 4 C, Fan-1.

119 'Pixie Variegata'

Species: *Bougainvillea x buttiana*

Parent: Mutant of 'Pixie'

Habit: Dwarf, growth moderately.

Vegetative Characters: Stem young coppery; Leaf blade 6.9 x 4.2 cm, ovate, variegated, speckled pattern, Yellow-green group 144 A, Fan-3, Yellow group 9 D, Fan-1; Thorn 1.5 cm long, straight to slightly curved.

Flower/Bract Characters: Flowering profuse, at middle of the branches (auxiliary); Bracts 2.0 x 1.3 cm, ovate, Red-purple group 72 C, Fan-2, non-persistent; Star 0.6 cm dia., Yellow group 4 C, Fan-1.

120 'Poultoni'

Species: *Bougainvillea x buttiana*

Parent: Seedling of *Bougainvillea x buttiana*

Habit: Tall, growth vigorous.

Vegetative Characters: Stem young green; Leaf blade 7.8 x 4.3 cm, elliptic, Green group 137 A, Fan-3; Thorn 1.3 cm long, straight.

Flower/Bract Characters: Flowering profuse, at the end of the branches; Bracts 4.3 x 3.3 cm, ovate, Red-purple group 70 B, Fan-2, non-persistent; Star 0.8 cm dia., Green-yellow group 1 C, Fan-1.

121 'Poultoni Special'

Species: *Bougainvillea x buttiana*

Parent: Hybrid seedling of *Bougainvillea x buttiana*

Habit: Tall, growth vigorous.

Vegetative Characters: Stem young green; Leaf blade 8.0 x 4.2 cm, ovate, Green group 137 A, Fan-3; Thorn 1.4 cm long, curved.

Flower/Bract Characters: Flowering profuse, at the end of the branches; Bracts 4.5 x 3.3 cm, ovate, Red-purple group 72 B, Fan-2, non-persistent; Star 0.8 cm dia., Green-yellow group 1 D, Fan-1.

122 'Pradhan's Profusion'

Species: *Bougainvillea spectabilis*

Parent: Hybrid Seedling of *Bougainvillea spectabilis*

Habit: Tall, growth vigorous.

Vegetative Characters: Stem young green; Leaf blade 7.9 x 4.6 cm, ovate, Green group 137 C, Fan-3; Thorn 3.0 cm long, curved.

Flower/Bract Characters: Flowering profuse, all along the branches; Bracts 3.4 x 1.9 cm, ovate, Red group 51 B, Fan-2, non-persistent; Star 0.7 cm dia., Green-yellow group 1 D, Fan-1.

123 'President Roosevelt'

Species: Bougainvillea peruviana

Parent: Unknown

Habit: Tall, growth vigorous.

Vegetative Characters: Stem young green; Leaf blade 7.8 x 5.1 cm, ovate, Green group 137 B, Fan-3; Thorn 1.1 cm long, curved.

Flower/Bract Characters: Flowering profuse, 3/4 of the branches; Bracts 4.1 x 3.0 cm, ovate, changing from Red group 47 A, Fan-1 to Red-purple group 60 C, Fan-2, non-persistent; Star 0.6 cm dia., Green-yellow group 1 C, Fan-1.

124 **'Profusion'**

Species: *Bougainvillea peruviana*

Parent: Unknown

Habit: Tall, growth vigorous.

Vegetative Characters: Stem young coppery; Leaf blade 7.0 x 4.5 cm, ovate, Green group 137 A, Fan-3; Thorn 1.5 cm long, straight.

Flower/Bract Characters: Flowering medium, 1/2 along the branches; Bracts 3.9 x 2.6 cm, elliptic, Red-purple group 72 B, Fan-2, non-persistent; Star 0.7 cm dia., Yellow group 2 D, Fan-1.

125 **'Red Glory'**

Species: *Bougainvillea spectabilis*

Parent: Seedling of *B. spectabilis*

Habit: Tall, growth vigorous.

Vegetative Characters: Stem young green; Leaf blade 6.8 x 4.7 cm, ovate, Yellow-green group 144 A, Fan-3; Thorn 1.0 cm long, slightly curved.

Flower/Bract Characters: Flowering profuse, all along the branches; Bracts 3.4 x 2.6 cm, ovate, Red-purple group 72 B, Fan-2, non-persistent; Star 0.7 cm dia., Green-yellow group 1 C, Fan-1.

126 **'Red Triangle'**

Species: *Bougainvillea spectabilis*

Parent: Unknown

Habit: Tall, growth vigorous.

Vegetative Characters: Stem young green; Leaf blade 7.2 x 4.7 cm, ovate, Green group 137 C, Fan-3, hairy; Thorn 1.4 cm long, curved.

Flower/Bract Characters: Flowering profuse, all along the branches; Bracts 3.6 x 2.2 cm, elliptic, Red-purple group 63 B, Fan-2, non-persistent; Star 0.7 cm dia., Green-yellow group 1 C, Fan-1.

127 **'Refulgens'**

Species: *Bougainvillea spectabilis*

Parent: Unkown

Habit: Tall, growth vigorous.

Vegetative Characters: Stem young green; Leaf blade 10.6 x 5.0 cm, elliptic, Green group 137 C, Fan-3; Thorn 2.3 cm long, curved.

Flower/Bract Characters: Flowering profuse, all along the branches; Bracts 3.6 x 2.1 cm, elliptic, Violet group 84 B, Fan-2, non-persistent; Star 0.7 cm dia., Yellow-green group 144 D, Fan-3.

128 **'Rhodamine'**

Species: *Bougainvillea x buttiana*

Parent: Unknown

Habit: Dwarf, growth medium.

Vegetative Characters: Stem young green; Leaf blade 7.4 x 4.1 cm, elliptic, Green group 137 C, Fan-3; Thorn 1.1 cm long, curved.

Flower/Bract Characters: Flowering medium, at the end of the branches; Bracts 4.2 x 3.0 cm, ovate, Red group 53 D, Fan-1, non-persistent; Star 0.6 cm dia., Green-yellow group 1 D, Fan-1.

129 'Rosa Multiflora'

Species: *Bougainvillea spectabilis* and *Bougainvillea x buttiana*

Parent: Hybrid of 'Mrs. Fraser' X 'Louis Wathen'

Habit: Tall, growth vigorous.

Vegetative Characters: Stem young greenish coppery; Leaf blade 8.9 x 4.0 cm, ovate to elliptic, Yellow-green group 144 A, Fan-3; Thorn 1.4 cm long, straight.

Flower/Bract Characters: Flowering profuse, all along the branches; Bracts 4.1 x 2.8 cm, ovate, Red-purple group 70 A, Fan-2, non-persistent; Star 1.1 cm dia., Yellow group 9 D, Fan-1.

130 'Rose Queen'

Species: *Bougainvillea x buttiana*

Parent: Hybrid seedling

Habit: Tall, growth moderately vigorous.

Vegetative Characters: Stem young coppery; Leaf blade 7.1 x 4.6 cm, ovate, Green group 137 A, Fan-3; Thorn 1.8 cm long, slightly curved.

Flower/Bract Characters: Flowering profuse, all along the branches; Bracts 3.6 x 2.7 cm, ovate, Red-purple group 70 A, Fan-2, non-persistent; Star 0.7 cm dia., Yellow group 2 D, Fan-1.

131 'Rosea'

Species: *Bougainvillea spectabilis*

Parent: Seedling of *Bougainvillea spectabilis*

Habit: Medium, drooping, growth intermediate.

Vegetative Characters: Stem young green; Leaf blade 7.2 x 4.3 cm, ovate, Green group 138 A, Fan-3; Thorn 1.4 cm long, straight.

Flower/Bract Characters: Flowering profuse, all along the branches; Bracts 3.8 x 2.3 cm, ovate, Red-purple group 72 B, Fan-2, non-persistent; Star 0.8 cm dia., Green-yellow group 2 D, Fan-1.

132 'Rosea Fuschea'

Species: *Bougainvillea peruviana*

Parent: Seedling of *Bougainvillea peruviana*

Habit: Medium, growth vigorous

Vegetative Characters: Stem young greenish coppery; Leaf blade 8.1 x 4.6 cm, ovate, Green group 137 A, Fan-3; Thorn 1.6 cm long, straight.

Flower/Bract Characters: Flowering profuse, at the end of the branches; Bracts 3.1 x 2.3 cm, ovate, Red-purple group 60 C, Fan-2, non-persistent; Star 0.7 cm dia., Yellow group 2 D, Fan-1.

133 'Roseville's Delight'

Species: *Bougainvillea x buttiana*

Parent: Bud sport of 'Mrs. McClean'

Habit: Medium, growth vigorous.

Vegetative Characters: Stem young greenish coppery; Leaf blade 6.1 x 3.9 cm, ovate, Green group 137 C, Fan-3; Thorn 1.5 cm long, straight.

Flower/Bract Characters: Flowering profuse, at the end of the branches; Bracts 2.2 x 1.1 cm, elliptic to ovate, change from Orange Red group 35 A, Fan-1 to Red-purple group 73 B, Fan-2, persistent; Floral tube absent.

134 **'Royal Daupline'**

Species: *Bougainvillea x buttiana*

Parent: Unknown

Habit: Tall, growth vigorous.

Vegetative Characters: Stem young greenish coppery; Leaf blade 8.5 x 6.5 cm, ovate, variegated, Green group 138 B, Fan-3, Yellow-green group 148 C, Fan-3; Thorn 1.0 cm long, straight.

Flower/Bract Characters: Flowering profuse, at the end of the branches; Bracts 4.1 x 3.4 cm, ovate, Red-purple group 71 B, Fan-2, non-persistent; Star 1.1 cm dia., Yellow group 4 D, Fan-1.

135 **'Ruarka'**

Species: *Bougainvillea x buttiana*

Parent: Seedling of *Bougainvillea x buttiana*

Habit: Tall, growth vigorous.

Vegetative Characters: Stem young coppery; Leaf blade 7.7 x 4.1 cm, ovate, Green group 138 A, Fan-3; Thorn 1.8 cm long, straight.

Flower/Bract Characters: Flowering profuse, along the branches; Bracts 4.0 x 2.8 cm, ovate, Red-purple group 70 B, Fan-2, non-persistent; Star 0.6 cm dia., Yellow group 2 D, Fan-1.

136 **'Sachidananda'**

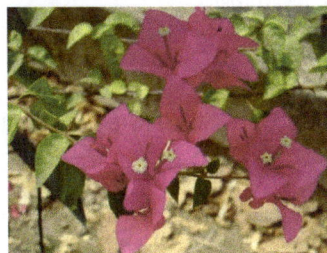

Species: *Bougainvillea glabra*

Parent: Seedling of 'Snow Queen'

Habit: Medium, growth vigorous.

Vegetative Characters: Stem young greenish coppery; Leaf blade 5.1 x 3.3 cm, ovate, Yellow-green group 144 A, Fan-3; Thorn 1.5 cm long, straight.

Flower/Bract Characters: Flowering profuse, all along the branches; Bracts 3.3 x 2.3 cm, elliptic, Red-purple group 70 D, Fan-2, persistent; Star 0.7 cm dia., Green-yellow group 1 C, Fan-1.

137 **'Scarlet Glory'**

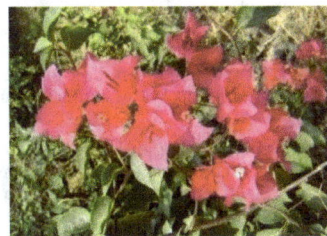

Species: *Bougainvillea x buttiana*

Parent: Bud sport of 'Padmi'

Habit: Medium, growth intermediate.

Vegetative Characters: Stem young coppery; Leaf blade 6.1 x 4.2 cm, broadly ovate, Green group 137 A, Fan-3; Thorn 1.0 cm long, curved.

Flower/Bract Characters: Flowering medium, at the end of the branches; Bracts 3.1 x 2.6 cm, ovate, Red-purple group 71 B, Fan-2, non-persistent; Star 0.5 cm dia., Green-yellow group 1 D, Fan-1.

138 **'Scarlet Queen'**

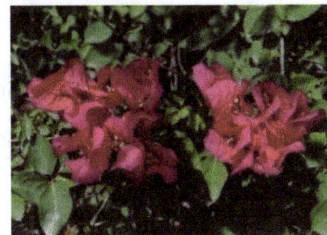

Species: *Bougainvillea x buttiana*

Parent: Bud sport of 'Mrs. Butt'

Habit: Medium, growth restricted.

Vegetative Characters: Stem young green; Leaf blade 6.7 x 4.5 cm, ovate, Green group 137 B, Fan-3; Thorn 1.0 cm long, straight.

Flower/Bract Characters: Flowering medium, at the end of the branches; Bracts 3.0 x 2.5 cm, elliptic, Red-purple group 69 B, Fan-2, non-persistent; Most of the flowers lack star, in some star 0.5 cm dia., Yellow group 2 C, Fan-1.

139 'Scarlet Queen Variegata'

Species: *Bougainvillea x buttiana*

Parent: Bud Sport of 'Scarlet Queen'

Habit: Tall, growth vigorous.

Vegetative Characters: Stem young coppery; Leaf blade 5.9 x 4.9 cm, broadly ovate, variegated (creamy white at margin), Green group 138 A, Fan-3, Yellow group 5 D, Fan-3; Thorn 1.2 cm long, straight.

Flower/Bract Characters: Flowering sparse, at the end of the branches; Bracts 2.6 x 2.2 cm, elliptic, Red-purple group 61 B, Fan-2, non-persistent; Star 0.4 cm dia., Orange group 27 B, Fan-1.

140 'Sensation'

Species: *Bougainvillea x buttiana*

Parent: Seedling of *Bougainvillea x buttiana*

Habit: Tall, growth vigorous.

Vegetative Characters: Stem young greenish coppery; Leaf blade 7.9 x 5.5 cm, ovate, Green group 137 A, Fan-3; Thorn 1.6 cm long, straight.

Flower/Bract Characters: Flowering profuse, 1/3 of the branches; Bracts 4.1 x 3.9 cm, ovate, Red-purple group 71 C, Fan-2, non-persistent; Star 0.7 cm dia., Yellow group 2 C, Fan-1.

141 'Shubhra'

Species: *Bougainvillea peruviana*

Parent: Bud sport of 'Mary Palmer'

Habit: Tall, growth vigorous.

Vegetative Characters: Stem young greenish coppery; Leaf blade 7.1 x 5.7 cm, ovate, Yellow-green group 143 A, Fan-3; Thorn 1.8 cm long, curved.

Flower/Bract Characters: Flowering profuse, all along the branches; Bracts 3.9 x 3.1 cm, ovate, Yellow White group 158 D, Fan 4, non-persistent; Star 0.8 cm dia., Yellow group 3 D, Fan-1.

142 'Shweta'

Species: *Bougainvillea glabra*

Parent: Bud sport of 'Trinidad'

Habit: Medium, drooping, growth restricted.

Vegetative Characters: Stem young green; Leaf blade 6.9 x 3.8 cm, elliptic, Yellow-green group 144 A, Fan-3; Thorn 1.0 cm long, curved.

Flower/Bract Characters: Flowering profuse, 1/4 to 3/4 of the branches; Bracts 3.1 x 2.0 cm, elliptic, Green White group 157 D, Fan 4, persistent; Star 0.6 cm dia., Green-yellow group 1 C, Fan-1.

143 'Silver Line'

Species: *Bougainvillea x buttiana*

Parent: Unknown

Habit: Dwarf, growth intermediate.

Vegetative Characters: Stem young greenish coppery; Leaf blade 6.9 x 4.6 cm, ovate, variegated (yellow at margins), Green group 138 A, Fan-3; Thorn 1.0 cm long, straight.

Flower/Bract Characters: Flowering profuse, at the end of the branches; Bracts 3.8 x 2.9 cm, ovate, Red-purple group 71 C, Fan-2, non-persistent; Star 0.6 cm dia., Yellow group 2 D, Fan-1.

144 **'Singapore Dark Red'**

Species: *Bougainvillea peruviana*

Parent: Unknown

Habit: Tall, growth vigorous.

Vegetative Characters: Stem young greenish coppery; Leaf blade 6.7 x 4.5 cm, ovate, Green group 137 C, Fan-3; Thorn 2.0 cm long, straight.

Flower/Bract Characters: Flowering medium, all along the branches; Bracts 3.8 x 3.1 cm, ovate, Red-purple group 69 B, Fan-2, non-persistent; Star 0.8 cm dia., Yellow group 2 D, Fan-1.

145 **'Sonnet'**

Species: *Bougainvillea x buttiana*

Parent: Seedling of *Bougainvillea x buttiana*

Habit: Medium, drooping, growth vigorous.

Vegetative Characters: Stem young greenish coppery; Leaf blade 6.5 x 4.4 cm, ovate, Green group 137 C, Fan-3; Thorn 1.4 cm long, straight.

Flower/Bract Characters: Flowering profuse, at the end of the branches; Bracts 3.7 x 2.6 cm, ovate, Red-purple group 72 B, Fan-2, non-persistent; Star 0.7 cm dia., Green-yellow group 1 D, Fan-1.

146 **'Sova'**

Species: *Bougainvillea glabra*

Parent: Unknown

Habit: Dwarf, growth intermediate.

Vegetative Characters: Stem young green; Leaf blade 4.4 x 2.0 cm, elliptic, Green group 137 C, Fan-3; Thorn 1.5 cm long, straight.

Flower/Bract Characters: Flowering profuse, all along the branches; Bracts 3.1 x 2.1 cm, ovate, White group 155 B, Fan 4, persistent; Star 0.7 cm dia., Yellow group 3 C, Fan-1.

147 **'Slendens'**

Species: *Bougainvillea spectabilis*

Parent: Seedling of *Bougainvillea spectabilis*

Habit: Tall, growth vigorous.

Vegetative Characters: Stem young green; Leaf blade 10.6 x 4.5 cm, elliptic, Green group 137 A, Fan-3, hairy; Thorn 2.7 cm long, curved.

Flower/Bract Characters: Flowering profuse, all along the branches; Bracts 4.6 x 3.2 cm, ovate, Red-purple group 72 B, Fan-2, non-persistent; Star 0.7 cm dia., Green-yellow group 1 D, Fan-1.

148 **'Spring Festival'**

Species: *Bougainvillea spectabilis X Bougainvillea x buttiana*

Parent: Hybrid seedling of 'Thomsaii' x 'Louise Wathen'

Habit: Tall, growth vigorous.

Vegetative Characters: Stem young greenish coppery; Leaf blade 8.3 x 3.7 cm, ovate, Green group 137 A, Fan-3; Thorn 1.9 cm long, straight.

Flower/Bract Characters: Flowering profuse, all along the branches; Bracts 5.0 x 3.2 cm, ovate, Red-purple group 72 B, Fan-2, non-persistent; Star 0.7 cm dia., Yellow group 3 D, Fan-1.

149 'Sri Durga"

Species: *Bougainvillea peruviana*

Parent: Unknown

Habit: Medium, growth intermediate.

Vegetative Characters: Stem young greenish coppery; Leaf blade 6.6 x 4.1 cm, ovate, Green group 137 D, Fan-3; Thorn 1.1 cm long, slightly curved to straight.

Flower/Bract Characters: Flowering profuse, 1/2 of the branches; Bracts 3.0 x 1.9 cm, ovate, Red-purple group 72 B, Fan-2, non-persistent; Star 0.6 cm dia., Yellow group 3 D, Fan-1.

150 'Srinivasa'

Species: *Bougainvillea x buttiana*

Parent: Seedling of 'Jayalakshmi'

Habit: Tall, growth vigorous.

Vegetative Characters: Stem young greenish coppery; Leaf blade 7.4 x 4.8 cm, ovate, Green group 137 B, Fan-3; Thorn 1.3 cm long, straight.

Flower/Bract Characters: Flowering profuse, at the end of the branches; Bracts 3.3 x 2.2 cm, ovate, Red-purple group 63 A, Fan-2, non-persistent; Star 0.7 cm dia., Green-yellow group 1 D, Fan-1.

151 'Stanza'

Species: *Bougainvillea glabra*

Parent: Seedling of *Bougainvillea glabra*

Habit: Medium, growth vigorous.

Vegetative Characters: Stem young coppery; Leaf blade 9.7 x 6.2 cm, ovate, Green group 137 B, Fan-3; Thorn 1.2 cm long, straight.

Flower/Bract Characters: Flowering profuse, all along the branches; Bracts 3.7 x 2.5 cm, elliptic, Red-purple group 71 B, Fan-2, non-persistent; Star 0.6 cm dia., Green-yellow group 1 D, Fan-1.

152 'SummerTime'

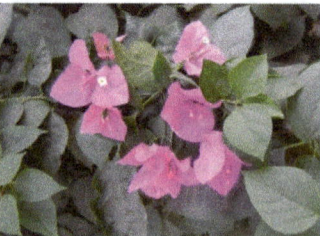

Species: *Bougainvillea spectabilis x Bougainvillea x buttiana*

Parent: Hybrid seedling of 'Thomsaii' x 'Louise Wathen'

Habit: Tall, growth vigorous.

Vegetative Characters: Stem young greenish coppery; Leaf blade 10.1 x 5.9 cm, ovate, Green group 137 A, Fan-3; Thorn 1.0 cm long, curved.

Flower/Bract Characters: Flowering medium, at the end of the branches; Bracts 3.1 x 2.1 cm, ovate, Red-purple group 70 B, Fan-2, non-persistent; Star 0.6 cm dia., Green-yellow group 1 D, Fan-1.

153 'Superba'

Species: *Bougainvillea x buttiana*

Parent: Unknown

Habit: Tall, growth vigorous

Vegetative Characters: Stem young green; Leaf blade 5.3 x 3.8 cm, ovate, Green group 137 C, Fan-3; Thorn 1.1 cm long, straight.

Flower/Bract Characters: Flowering profuse, at the end of the branches; Bracts 4.4 x 3.3 cm, elliptic, Red-purple group 72 A, Fan-2, non-persistent; Star 0.7 cm dia., Yellow group 2 D, Fan-1.

154 **'Suvarna'**

Species: *Bougainvillea glabra*

Parent: Unknown

Habit: Medium, growth vigorous.

Vegetative Characters: Stem young green; Leaf blade 7.0 x 3.8 cm, elliptic, Green group 137 C, Fan-3; Thorn 1.1 cm long, slightly curved.

Flower/Bract Characters: Flowering profuse, all along the branches; Bracts 3.7 x 2.8 cm, ovate, Yellow Orange group 19 B, Fan-1, non-persistent; Star 0.6 cm dia., Yellow group 5 D, Fan-1.

155 **'Sydney'**

Species: *Bougainvillea glabra*

Parent: Unknown

Habit: Medium, growth vigorous.

Vegetative Characters: Stem young green; Leaf blade 6.7 x 3.7 cm, elliptic, Green group 137 C, Fan-3; Thorn 1.2 cm long, slightly curved.

Flower/Bract Characters: Flowering profuse, all along the branches; Bracts 3.8 x 2.5 cm, elliptic, Red-purple group 72 B, Fan-2, persistent; Star 0.7 cm dia., Yellow group 2 C, Fan-1.

156 **'Tetra Mrs. McClean'**

Species: *Bougainvillea x buttiana*

Parent: Tetraploid of 'Mrs. McClean'

Habit: Tall, growth vigorous.

Vegetative Characters: Stem young greenish coppery; Leaf blade 6.8 x 5.3 cm, broadly ovate, Green group 137 A, Fan-3; Thorn 0.8 cm long, straight.

Flower/Bract Characters: Flowering medium, at the end of the branches; Bracts 3.2 x 2.9 cm, ovate, changing Orange group 26 A, Fan-1 to Red-purple group 62 A, Fan-2, non-persistent; Star 0.7 cm dia., Yellow group 2 D, Fan-1.

157 **'Texas Dawn'**

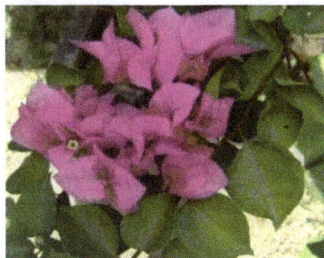

Species: *Bougainvillea x buttiana*

Parent: Unknown

Habit: Medium, growth vigorous.

Vegetative Characters: Stem young coppery; Leaf blade 6.4 x 4.1 cm, ovate, Green group 137 C, Fan-3; Thorn 0.8 cm long, straight.

Flower/Bract Characters: Flowering profuse, at the end of the branches; Bracts 3.3 x 2.5 cm, ovate, Red-purple group 70 B, Fan-2, non-persistent; Star 0.6 cm dia., Green-yellow group 1 C, Fan-1.

158 **'Theresa Jacobs'**

Species: *Bougainvillea peruviana*

Parent: Unknown

Habit: Tall, growth vigorous.

Vegetative Characters: Stem young coppery; Leaf blade 6.9 x 4.6 cm, ovate, Yellow-green group 148 A, Fan-3; Thorn 1.6 cm long, straight.

Flower/Bract Characters: Flowering profuse, half of the branches; Bracts 3.6 x 2.7 cm, ovate, Red-purple group 70 A, Fan-2, non-persistent; Star 0.7 cm dia., Yellow group 6 D, Fan-1.

159 'Thimma'

Species: *Bougainvillea peruviana*

Parent: Bud sport of 'Mary Palmer'

Habit: Tall growth vigorous.

Vegetative Characters: Stem young yellow; Leaf blade 7.2 x 4.8 cm, ovate, variegated (yellow in middle), Yellow-green group 144 A, Fan-3; Thorn 1.4 cm long, slightly curved.

Flower/Bract Characters: Flowering profuse, all along the branches; Bracts 4.8 x 3.5 cm, ovate, three type of bract colour - Red-purple group 72 B, Fan-2, Orange-white group 159 B, Fan 4, Orange-white group 159 B, Fan 4, with Red Purple group 72 B, Fan-2, non-persistent; Star 0.7 cm dia., Green-yellow group 1 D, Fan-1.

160 'Thimma Sport'

Species: *Bougainvillea peruviana*

Parent: Unknown

Habit: Tall, growth vigorous.

Vegetative Characters: Stem young greenish coppery; Leaf blade 5.9 x 3.5 cm, ovate, Green group 137 B, Fan-3; Thorn 1.0 cm long, curved.

Flower/Bract Characters: Flowering profuse, Bracts borne all along the branches; Bracts 4.2 x 2.7 cm, ovate, Red-purple group 73 C, Fan-2, non-persistent; Star 0.7 cm dia., Yellow group 2 C, Fan-1.

161 'Tomato Red'

Species: *Bougainvillea spectabilis*

Parent: Unknown

Habit: Medium, growth intermediate.

Vegetative Characters: Stem young green; Leaf blade 6.9 x 4.6 cm, elliptic, Green group 137 B, Fan-3; Thorn 1.0 cm long, curved.

Flower/Bract Characters: Flowering profuse, Bracts borne all along the branches; Bracts 3.7 x 2.6 cm, elliptic, Red group 54 A, Fan-1, non-persistent; Star 0.7 cm dia., Green-yellow group 1 D, Fan-1.

162 'Versicolour'

Species: *Bougainvillea x buttiana*

Parent: Hybrid seedling

Habit: Tall, growth vigorous.

Vegetative Characters: Stem young greenish coppery; Leaf blade 5.8 x 4.5 cm, ovate, Green group 139 A, Fan-3; Thorn 1.0 cm long, straight.

Flower/Bract Characters: Flowering profuse, at the end of the branches; Bracts 3.5 x 2.9 cm, ovate, Red-purple group 70 A, Fan-2, non-persistent; Star 0.6 cm dia., Green-yellow group 1 D, Fan-1.

163 'Vishakha'

Species: *Bougainvillea peruviana*

Parent: Bud sport of 'Mrs. H. C. Buck'

Habit: Medium, growth vigorous.

Vegetative Characters: Stem young green; Leaf blade 8.0 x 4.1 cm, elliptic, Green group 137 D, Fan-3; Thorn 1.2 cm long, straight.

Flower/Bract Characters: Flowering medium and free, all along the branches; Bracts 3.2 x 1.9 cm, ovate, Red-purple group 71 B, Fan-2, non-persistent; Star 0.5 cm dia., Green-yellow group 1 C, Fan-1.

164 **'VishakaVariegata'**

Species: *Bougainvillea peruviana*

Parent: Mutant of 'Vishakha'

Habit: Medium, growth vigorous.

Vegetative Characters: Stem young green; Leaf blade 7.2 x 4.0 cm, elliptic, variegated (white margin), Yellow-green group 147 B, Fan-3; Thorn 1.2 cm long, straight.

Flower/Bract Characters: Flowering sparse and free, all along the branches; Bracts 3.0 x 1.8 cm, ovate, Red-purple group 69 B, Fan-2, non-persistent; Star 0.7 cm dia., Green-yellow group 1 C, Fan-1.

165 **'Vitthal Variegata'**

Species: *Bougainvillea spectabilis*

Parent: Unknown

Habit: Medium, growth vigorous

Vegetative Characters: Stem young greenish coopery; leaf blade 7.2 x 4.0 cm, medium ovate, hairy, variegated, speckled type of variegation, Yellow-green group 144A, Fan 3, Yellow group 9D, Fan -1; Thorn 1.0 cm long, straight.

Flower/Bract Characters: Flowering sparse and free, at the end of branches; Bracts 3.0 x 1.8 cm, ovate, Red-purple group 72B, Fan-2, non-persistent; Star 0.7 cm dia., Green-yellow group 1C, Fan-1.

166 **'Wajid Ali Shah'**

Species: *Bougainvillea peruviana* and *B. spectabilis*

Parent: Hybrid seedling of 'Dr. B.P. Pal' x 'Mrs. Chico'

Habit: Medium, drooping, growth restricted.

Vegetative Characters: Stem young greenish coopery; leaf blade 7.8 x 5.2 cm, elliptic, Green group 137 C; Thorn 1.5 cm long, slightly curved.

Flower/Bract Characters: Flowering profuse, all along the branches; Bracts 3.9 x 3.0 cm, ovate, Red-purple group 70 B, Fan-2, non-persistent; Star 0.7 cm dia., Green-yellow group 1 D, Fan-1.

167 **'Yellow Queen'**

Species: *Bougainvillea x buttiana*

Parent: Bud sport of 'Mrs. McClean'

Habit: Tall, growth vigorous.

Vegetative Characters: Stem young green; Leaf blade 7.2 x 4.3 cm, ovate, Green group 137 A, Fan-3; Thorn 1.3 cm long, slightly curved.

Flower/Bract Characters: Flowering profuse, bracts borne all along the branches; Bracts 2.9 x 2.3 cm, ovate, Yellow group 10 B, Fan-1, non-persistent; Star 0.6 cm dia., Green-yellow group 1 D, Fan-1.

168 **'Zakariana'**

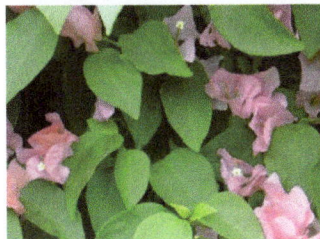

Species: *Bougainvillea spectabilis*

Parent: Unknown

Habit: Medium, growth intermediate.

Vegetative Characters: Stem young green; Leaf blade 9.1 x 5.5 cm, ovate, Green group 137 C, Fan-3; Thorn 2.0 cm long, slightly curved.

Flower/Bract Characters: Flowering profuse, all along the branches; Bracts 3.3 x 2.5 cm, ovate, Red group 51 C, Fan-1, non-persistent; Star 0.6 cm dia., Yellow group 2 C, Fan-1.

169 **'Zinna Barat'**

Species: _Bougainvillea glabra_

Parent: Seedling of _Bougainvillea glabra_

Habit: Medium, growth vigorous.

Vegetative Characters: Stem young green; Leaf blade 6.6 x 2.6 cm, elliptic, Yellow-green group 144 A, Fan-3; Thorn 1.1 cm long, curved.

Flower/Bract Characters: Flowering profuse, all along the branches; Bracts 3.8 x 2.0 cm, elliptic, Purple group 75 A, Fan-2, persistent; Star 0.7 cm dia., Green-yellow group 1 C, Fan-1.

170 **'ZakarianaVariegata'**

Species: _Bougainvillea spectabilis_

Parent: Mutant of 'Zakariana'

Habit: Medium, growth intermediate.

Vegetative Characters: Stem young green; Leaf blade 10.0 x 6.2 cm, ovate, variegated (light green in speckled pattern), Green group 137 C, Fan-3; Thorn 2.0 cm long, slightly curved.

Flower/Bract Characters: Flowering profuse, all along the branches; Bracts 3.5 x 2.8 cm, ovate, Red group 48 B, Fan-1, non-persistent; Star 0.6 cm dia., Yellow group 2 D, Fan-1.

171 **'Zulu Queen'**

Species: _Bougainvillea x buttiana_

Parent: Seedling of _Bougainvillea x buttiana_

Habit: Tall, growth moderate

Vegetative Characters: Stem young greenish coppery; Leaf blade 7.0 x 4.3 cm, ovate to elliptic, Green group 137 A, Fan-3; Thorn 1.3 cm long, straight

Flower/Bract Characters: Flowering profuse, at 1/3 of the branches; Bracts 3.8 x 2.9 cm, ovate, Red-purple group 72 A, Fan-2, non-persistent; Star 0.7 cm dia., Yellow group 3 C, Fan-1

SOME OTHER VARIETIES

'Brasiliensis'

'Easter Parade'

'Formosa'

'Texas King'

'Espinosa'

'Afterglow' (Salmon)

'Hugh Evans' (Coral)

'Rosenka' (Orange)

'Mrs. Eva Variegata'

'Millionaire' / 'Snow White'

'Aiskrim' ('Surprise', /'Miss Universe'),

'Orange Ice Variegata'

'Royal Purple'

'Tropical Bouquet'

'Smarty Pants'

'Orange Ice'

'White Stripe'

'Bengal Orange'

'Raspberry Ice'

'Scarlet –O-Hara'

'Sundown Orange'

'Tequila Sunrise'

'James Walker'

'Pink Pixie'

'Rosa Preciosa'

'Temple Fire'

'Miami Pink'

'New River'

'Jubliee Showlady'

'Maureen Hatten'

'Sanderiana'

'Twilight Delight'

'Crimson Red'

'Don Fernando'

'Juanita Hatten'

'La Jola'

'Queen Violet'

'Sao Paulo'

'Isla Maroda'

'Key West Alba'

'California Gold'

'Marilyn-Hatten'

'Alisia'

'Beba Pink'

'Fifina' (Purple)

'Yuyu' (Light Blue)

'Carpet Pink'

'Carpet Purple'

'Carpet white'

'Bridal Bouquet'

'Helen Jhonson' Pink

'Ole' (Reddish)

'Scarlet-O-Hara' (Red)

'Snow Purple' (Rosy & Purple)

'Coconut Ice' (Lilac & White)

'Flamingo' (Pink White)

'Surprise' (Lilac & White)

'Yani's Delight' (Pink White)

'Blue Berry Ice' (Purple)

'Delta Dawn' (Yellow)

'Double Delight' (Pinky White)

'Fantasy Red'

'Orange Ice'

'Picta Aurea'

'Pixie Queen

'Royal Bengal Red'

'Royal Dauphine' (Red)

'Rubra (Red)

'Vicky Thimma' (Lilac & White)

'White Stripe'

'Sunvillea Cream'

'Sunvillea Pink'

'Sunvillea Rose'

'Java White'

'New River'

'Josephine-Yuyu'

'Rosenka'

'Twilight Delight'

'Sundown Orange'

'Fafina'

'Josephine Ali'

'Carpet Purple'

NEW VARIETIES OF BOUGAINVILLEA

References

Alfieri, S.A. 1970. *Bougainvillea*, a new disease caused by *Phytophthora parasitica*. *Phytopathology*, 60:1806-1808.

Anonymous. 1854. *Bougainvillea spectabilis. Curtis's Bot.* Mag. 4810

Anonymous. 1860. *Bougainvillea spectabilis. Gardeners' Chronica*. 19:431

Anonymous. 1894a. *Bougainvillea glabra* var. 'Sanderiana'. *Proceeding of Royal Horticultural Society*. 17:8.

Anonymous. 1894b. *Bougainvillea glabra* var. 'Penang'. *Proceeding of RoyalHorticulture Society*. 17:84.

Anonymous. 1900. *Bougainvillea spectabilis* var. 'Maud Chettleburgh'. *Proceeding of Royal Horticultural Society*. 25:83.

Anonymous. 1904. *Bougainville aspectabilis* var. 'Canellii'. *Gardeners' Chronica*. 36:383.

Anonymous. 1910. 'Rosa Catalina'. *Proceedings of Royal Horticultural Society*. 35:59.

Anonymous. 1919. 'Lady Watts'. Rep. agric. Dep. St Vincent, Peru.

Anonymous. 1923. 'Mrs. Butt'. *Proc. R. Hort. Soc.* 48:46

Anonymous. 1924. *Bougainvillea* 'Mrs. Butt'. *Kew Bull.* 8:335

Anonymous. 1932. 'Mrs. Fraser'. *Gardeners' Chronica*. (3rd ser.) 91: 11

Anonymous. 1934. 'Mrs. Lancaster'. *Gardeners' Chronica*. 95:385

Anonymous. 1951. 'Mary Palmer'. *J. R. Hort. Soc.* 76:238

Anonymous. 1961. New varieties of Bougainvilleas. *Indian Horticulture*. 5(3):11

Anonymous. 2015 R.H.S. Colour Chart (6th Edition), The Royal Horticultural Society Media, London, U.K. in association with the Flower Council of Holland .

Arya, H.C. 1963. A New Colletotrichum On *Bougainvillea*. *Indian Phytopathology*. 16(2):234-7.

Bailey, L.H. 1942. The Standard Cyclopedia of Horticulture. The MacMillan Company, New York, 1: 533-34.

Banerji, B.K. 2007. Multibracted *Bougainvillea*. *In* Hi-Tech Floriculture. (Eds.) Satyanarayan Reddy, B.S., Janakiram, T., Balaji, Kulkarni, S. and Misra, R.L. *Indian Society of Ornamental Horticulture*, IARI, New Delhi . pp: 190- 192.

Banerji, B.K. 2008. Mutation and mysterious world of multibracted *Bougainvillea*. *Floriculture Today*, 15: 12-15.

Banerji, B.K. and Datta, S.K. 1986. 'Improvement of ornamental plants by induced mutation: *Bougainvillea*. *Floriculture Today*, 3: 4-9.

Banerji, B.K. and Datta, S.K. 1987. Gamma ray induced chlorophyll variegated mutants in *Bougainvillea* cv. 'Los Banos Beauty'. *Journal of Nuclear Agricultural Biology*, 16(1): 48-50.

Banerji, B.K. and Datta, S.K. 1987. Gamma ray induced genetic improvement of double bracted *Bougainvillea*. Paper presented at All India Seminar on 'Advances in Botanical Research in India during the last ten years'. Nov. 1-3, 1987, Dept. of Botany, Dungar College, Bikaner, Rajasthan.

Banerji, B.K. and Datta, S.K. 1987. Gamma ray induced mutation in double bracted *Bougainvillea* cv. 'Los Banos Beauty'. *Journal of Nuclear Agricultural Biology*, 16: 48-50.

Banerji, B.K. and Datta, S.K. 1988. A review on variegated *Bougainvillea*. *Floriculture Today*, 8: 1-6.

Banerji, B.K., Nath, P. and Datta, S.K. 1987. Mutation breeding in double bracted *Bougainvillea* cv. 'Roseville Delight'. *Journal of Nuclear Agricultural Biology*, 16(1): 45-47.

Banerji, B.K., Nath, P. and Gupta, M.N. 1983. Gamma ray induced somatic mutation in *Bougainvillea* cv. 'Roseville's Delight' I: Induction of chlorophyll   mutation. In: 'National Symposium on Recent Development in Nuclear and Allied Techniques and their application in Agriculture, Biology & Animal Science, May 5- 7, G .B. Pant Univ. of Agric. & Tech., Pantnagar, pp. 41.

Banerji, B.K. and Datta, S.K. 1967. A natural triploid in *Bougainvillea*. *Indian Journal of Horticulture*. 24:106-8.

Banerji, B.K. and Dwivedi, A.K. 2013. Bougainvillea cultivars evolved as seedling from 'Mrs. Butt' – A review. 2009. Ind. J. Hort. 8 (2) : 45-47.

Bor, N.L. andRaizada, M. B. 1954. Some beautiful Indian climbers and shrubs, 286 pp. The Bombay Natural History Society, Bombay: pp. 267-74.

Bose, T.K. andHore, B. 1967. Effect of growth retardants on some varieties of *Bougainvillea*. *Science Culture*, 33(18):379-80.

Chandra, R. 1961. Budding in Bougainvilleas. *Indian Horticulture*, 5(4):18-9.

Chatterjee, J., Mondal, A.A. and Datta, S.K. 2007.Using of RAPD analysis to determine genetic diversity and relationship among Bougainvillea cultivars at intra and inter-specific levels. *Horticulture environment and biotechnology*, 48 (1): 43-51.

Choudhary, B. and Singh, B. 1981. The International *Bougainvillea* Checklist. Indian Agriculture Research Institute., New Delhi.

Cooper, D.C. 1929. The Chromosomes in *Bougainvillea*. *Proceeding. National Academic Science*, 15:885-7.

Cooper, D.C. 1931. Microsporogenesis in *Bougainvillea glabra*. *American Journal of Botany* 18:337-58.

Costa, A.S. and Carvalho, A.M.B. 1960. Marginal leaf-curl in *Bougainvillea* caused by Mites. *Bragantia* 19 (Suppl.):134-40.

Datta, S.K. and Banerji, B.K. 1997. Improvement of double bracted *Bougainvillea* through gamma ray induced mutation. *Frontiers in Plant Science*, 64: 395-400 (Editor Dr. Irfan A. Khan).

Datta, S.K. and Banerji, B.K. 1990. 'Los Banos Variegata' – A new double bracted chlorophyll variegate *Bougainvillea* induced by gamma rays. *Journal of Nuclear Agricultural Biology*, 19: 134-36.

Datta, S.K. and Banerji, B.K. 1994. Improvement of double bracted *Bougainvillea* through induced mutation. National Symposium on Frontiers in Plant Science Research, N.S.A. Khan Center of P.G. Studies and Research, Hyderabad, Feb. 13-14.

Datta, S.K. and Banerji, B.K. 1994. 'Mahara Variegata'- A new mutant of *Bougainvillea*. *Journal of Nuclear Agricultural Biology*, 23(2): 114-116.

Datta, S.K., Dwivedi, A.K. and Banerji, B.K. 1995. Investigations on gamma ray induced chlorophyll variegated mutants. *Journal of Nuclear Agricultural Biology*, 24: 237-47.

De Jussieu A.L. 1789. *Bougainvillea* Commer. In : Genera Plantarum., Whelden and Wesley Ltd., New York, 91.

Even-Chen, Z., Roy M. Sachs and Hackett, W.P. 1979. Control of flowering in Bougainvillea 'San Diego Red'. *Plant Physiol.* 64 : 646-651.

Germek, E. 1954. Hybridization of *Bougainvillea*. *Bragantia* 13:5-7.

Golby, Eric V. 1970. History of *Bougainvillea* in Florida (PP. 138-45), *In* Edwin A. M. Flowering vines of the world: An Encyclopedia of Climbing Plants. Hearthside Press Inc., Publishers, New York.

Gopalaswamiengar, K.S. 1935. Complete Gardening In India, The Hoslai Press, Bangalore: 259-62.

Gupta VN and Kher MA 1991. A note on the influence of auxins in regeneration of roots in the tip cuttings of *Bougainvillea* sp. var. Garnet Glory under intermittent mist. *Haryana J. Horticult. Sci.,* 20: 85-87.

Gupta, M N. 1972. Mutation breeding of some vegetatively propagated ornamentals. In: Progress in Plant Research, NBRI Silver Jubilee Publication, Today & Tomorrow's Printers & Publishers, New Delhi, pp 75.

Hackett, W.P. and Sachs, R.M. 1964. Flowering in *Bougainvillea*. *American Journal of Botany* 51: 663 (Abstract).

Hackett, W.P. and Sachs, R.M. 1965. Factors affecting flowering of *Bougainvillea*. *California Agriculture* 19(9): 13.

Hackett, W.P. and Sachs, R.M. 1966. Factors affecting flowering of *Bougainvillea* 'San Diego'. *Proceedings of the American Society for Horticultural Science*, 88: 606-12.

Hackett, W.P. and Sachs, R.M. 1967. Chemical control of flowering in *Bougainvillea* 'San Diego Red'. *Proceedings of the American Society for Horticultural Science,* 90: 361-4.

Hackett, W.P. and Sachs, R.M. 1968. Experimental separation of inflorescence of development from initiation in *Bougainvillea. Proceedings of the American Society for Horticultural Sci.,* 92: 615-21.

Hackett, W.P., Sachs, R.M. and DeBie, J. 1972. Growing *Bougainvillea* as a flowering pot plant. *California Agriculture* 26 (8): 12-13.

Hammad, I. 2009. Genetic variation among *Bougainvillea glabra* cultivars (Nyctagenaceae) detected by RAPD markers and isozymes patterns. *Res. J. Agril. Bio. Sci.,* 5(1):63-71.

Heimerl, A. 1900. Monographie der Nictaginaceen – Bougainvillea, Phaeoptilum, Colignonia, Denkschriften der Kaiserlichen Akademie der Wissenschaften,; Mathematisch-*Naturwissenschaftlichen Classe,* 70 : 97-137.

Holttum, R.E. 1938. The cultivated Bougainvilleas. *Gardeners' Chronica.* 103: 164-5.

Holttum, R.E. 1955b. The cultivated Bougainvilleas. II Malay. *Malayan Agri-Horticultural Association Magazine,* 12(2): 2-10.

Holttum, R.E. 1955a. The cultivated Bougainvilleas. I. Malay. *Malayan Agri-Horticultural Association Magazine,* 12(2): 2-10.

Holttum, R.E. 1955c. The cultivated Bougainvilleas. III. The varieties of *B. glabra. Malayan Agri-Horticultural Association Magazine,* 13(1): 13-22.

Holttum, R.E. 1957a. *Bougainvillea* 'Mary Palmer'. *Malayan Agri-Horticultural Association Magazine,* 14(1): 13.

Holttum, R.E. 1957b. Further notes on *Bougainvillea. Malayan Agri-Horticultural Association Magazine,* 14(2): 58.

Iredell, J. 1990. The *Bougainvillea* Growers Handbook. Simon & Sehusyer, Australia.

Iredell, J. 1995. Growing Bougainvilleas . Cassell Publishers Limited, London,

Jain, R.K. 1959. Some tetraploid variations in the inflorescences of *Bougainvillea glabra* Choisy. *Current Science,* 28: 21-2.

Jayanti, R. and Datta, S.K. 2006. Improvement of *Bougainvillea* varieties using Chemical Mutagens. In: 'National Conference on *Bougainvillea*', April12-13, 2006, NBRI, Lucknow, pp17.

Jayanti, R. Datta, S.K. and Verma, J.P. 1999. Effect of gamma rays on single bracted *Bougainvillea. Journal of Nuclear Agricultural Biology,* 28: 228-33.

Joiner, J.N. and Dickey, R.D. 1963-64. Some factors affecting flowering and propagation of *Bougainvillea glabra* Sander. *Proc. Fla St. hort. Soc.* 76:441-4.

Khoshoo, T.N. 1971a. Taxonomy of cultivated plants: Principles, procedures and prospects. *Sci. Cult.,* 37 : 313-15.

Khoshoo, T.N. 1971b. Hortorium taxonomy. *Indian Journal of Genetics and Plant Breeding,* 31(2): 305-315.

Khoshoo, T.N. 1990. Creativity in ornamentals. *Indian Horticulture,* 34: 25-29 & 33.

Khoshoo, T.N. and Zadoo, S.N. 1969. New perspectives in *Bougainvillea* breeding, *Journal of Heredity*, 60(6): 357-60.

Khoshoo, T.N. and Zadoo, S.N. 1986. Perspectives in *Bougainvillea* breeding Newsletter, *Bougainvillea* Society of India, 6(2) : 7-10.

Kochhar, V.K. and Ohri, D. 1977. Bio-chemical analysis of bract mutations in Bougainvilleas - I: 'H.C. Buck' family. *Z. Pflarizenzuchiq*, 79: 47-51.

Kochhar, V.K., Kochhar, S. and Ohri, D. 1979. Bio-chemical analysis of bract mutation in *Bougainvillea* cv. Mrs. H.C. Buck. *Current Science*, 48: 163-64.

Kumar, P.P., Jankiram, T., Bhat, K.V., Jain, R., Prasad, K.V. and Prabhu, K.V. 2014. Molecular characterization and cultivar identification in Bougainvillea spp. *Indian J. Agril. Sci.* 84 (8):1024-30.

Kumar, Satish, Roy, R.K., and Shrama, Girdhari. 2012 Bougainvilleas in Home Gardens. *Indian Bougainvillea Annual*, 24: 14-16.

Lancaster, S.P. 1951. The white *Bougainvillea* in India. *J. R. Hort. Soc.* 76(8) : 278.

Lancaster, S.P.1959. *Bougainvillea*. Bull. 41, National Botanical Gardens, Lucknow, 31 pp.

Leonardi, C. and Romano, D. 2003. Characteristics of Bougainvillea types widespread in the east of Sicily. *Italus Hortus*, 10(4):234-237.

Mac Daniels, L.H. 1981. A study of cultivars in *Bougainvillea* (Nyctanginaceae) *Baileya*, 21(2): 77-100.

Marigowda, M. H. 1960. Short notes on Bougainvilleas in Lal Bagh. *Lal Bagh Journal of Mysore Horticultural Society*, 5: 25-8.

Marigowda, M. H. 1960. Three outstanding varieties of *Bougainvillea* in Lal Bagh. *Lal Bagh Journal of Mysore Horticultural Society*, 5(3): 8-11.

Mukhopadhaya, D.P. and Bose, T.K. 1966. Improvement in the method of vegetative propagation in some varieties of Bougainvilleas. *Indian Journal of Horticulture*. 23: 185-6.

Mukhopadhaya, S. and Lakshmanan, V. 1957. A note on seed setting in *Bougainvillea*. S. *Indian Horticulture*, 5: 35-7.

Nair, P.K.K. 1965. Significance of pollen morphology in the study of cultivated plants. *Indian Agric*. 9: 53-87.

Nath, P., Banerji, B.K. and Gupta, M.N. 1983. Spontaneous and induced mutations in *Bougainvillea*. News Letter, *Bougainvillea* Festival, *Bougainvillea Society of India*. pp.1-6.

Ninan, T., Singh, M.P. and Swaminathan, M.S. 1959. Meiotic behaviour and pollen fertility in some varieties of *Bougainvillea*. *Journal of Indian Botanical Society* 38: 140-5.

Ohri, D. 1979. Cytogenetics of Garden Gladiolus and *Bougainvillea*. Ph.D. Thesis, Punjab University, Chandigarh.

Ohri, D. 1995. *Bougainvillea* Breeding. *In*: Advances in Horticulture Vol. 12 - Ornamental Plants, (Eds. K.L. Chadha & S.K. Bhattacharjee). Malhotra Publishing House, New Delhi, pp. 363-76.

Ohri, D. and Khoshoo, T.N. 1982. Cytogenetics of cultivated Bougainvilleas - X: Nuclear DNA content. *Z. Pjlanzenzuchtg*, 88: 168-73.

Ohri, D. and Zadoo, S.N. 1975. Cross-compatibility studies in some *Bougainvillea* varieties. *Incompatibility Newsletter*, 6: 35-39.

Ohri, D. and Zadoo, S.N. 1979. Cytogenetics of cultivated Bougainvilleas - VIII: Cross-compabtlity relation-ships and origin of *Bougainvillea* 'H.C. Buck' family. *Z. PjIanzenzuchtg*, 82: 182-86.

Ohri, D. and Zadoo, S.N. 1986, 97. Cytogenetics of cultivated *Bougainvillea* - IX: Precocious centromere division and origin of polyploid taxa. *Plant Breeding*, 227-31.

Pal, B.P. 1959. Introducing four new Bougainvilleas. *Indian Horticulture*, 4(1):16-22.

Pal, B.P. 1960. Beautiful Climbers in lndia. Indian Council of Agricultural Research, New Delhi (*Bougainvillea*), pp. 35-47.

Pal, B.P. and Swarup, V. 1974. Bougainvilleas. I.C.A.R., New Delhi.

Pancho, J.V. 1963. Notes on two outstanding new varieties of *Bougainvillea* in the Philippines. *Lal Bagh J. Mysore Hort. Soc.* 8(3) : 25-6.

Pancho, J.V. 967. Notes on double-bracted *Bougainvillea* in the Philippines. *Lal Bagh J. Mysore Hort. Soc.* 12(2) : 30-2.

Pancho, J.V. and Bardenas, A. E. 1959. Bougainvilleas in the Philippines. Baileya 7(3) : 91-100.

Pancho, J.V. and Capinpin, J. M. 1961. Haploidy in *Bougainvillea. Philippines Agriculture*, 45: 88-94.

Pancho, J.V., Capinpin, J. M. and Bardenas, A. E. 1960. Chromosome number and fertility tests in *Bougainvillea. Philippines Agriculture*, 45: 11-8.

Parsons, T. H. 1935. 'Mrs. Butt' *Bougainvillea* at Peradeniya. Trop. Agric. Mag. Ceylon agric. Soc. 85: 335.

Piattelli, M. and Imperato, F. 1970. Betacyamins from *Bougainvillea*. Phytochemistry 9: 455-8.

Pillal, P.K. 1963. A new technique on rooting of plant cuttings using growth regulators. *Madras agric, J.* 50: 29-30.

Popenoe, J. 1961. *Bougainvillea* culture. *Am. Hort, Mag.* 40(4): 319-24.

Ramina, A., Hackett, W.P., and Sach, R.M. 1979. Flowering in Bougainvillea – A Function of Assimilate and Nutrient Diversion, *Plant Physio.* 64:810-813.

Rao, U. N. and Muthuswamy, S. 1955. The white *Bougainvillea. S. Indian Hort.* 3: 114-5.

Rastogi, R.R., Singh, S., Sharma, G. and Roy, R.K. 2015. Splash Colour in the Landscape by Bougainvilleas. *Indian Bougainvillea Annual*, 26:23-26.

Roy, R. K., Banerji, B.K. and Goel A.K. 2007. Bougainvillea – Germplasm collections at NBRI. *Indian Bougainvillea Annual* (20):10-13.

Roy, R. K., Kumar, Satish and Sindhu, S.S. 2012. History of Migration of Bougainvilleas and Development of Indian Bred Cultivars. *Indian Bougainvillea Annual*, 24: 04-13.

Roy, R.K. 1987. *Bougainvillea* in landscaping. Flower Show Bull., Calcutta Flower Growers' Asso., 21.

Roy, R.K. 2008. Bougainvillea – An excellent ornamental for every garden. *Bougainvillea Annual* 21: 18-20.

Roy, R.K. 2009. Bougainvillea Nursery: A successful commercial entity for semi-urban areas. *Bougainvillea Annual* 22: 1- 6.

Roy, R.K. 2010. New pruning techniques in bougainvilleas for floriferous flowering. *Bougainvillea Annual,* 23: 36-38.

Roy, R.K. 2010. Splash colour to the city landscape by Bougainvilleas. *Bougainvillea Annual,* 23: 16-17

Roy, R.K. and Singh, Shilpi. 2010. Migration and domestication of Bougainvillea: a historical review. Chronica, Int'l Soc. Hort. Sci.:20-30.

Roy, R.K. 2013. Fundamentals of Garden Designing, New India Publishing House, New Delhi, p.1-699.

Roy, R.K. and Sharma, S.C. 2002. *Bougainvillea*: The charming beauty. *Hort. J.,* 12(1-4): 25-31.

Roy, R.K., Goel, A.K., Singh, S. and Prasad, R. 2014. Characterization of new cultivars of ornamentals. Classical and Methods in Plant Taxonomy and Biosystematics, pp 477-483.

Roy, R.K., Prasad, R. & Singh, S. 2013. Splash a colourful touch in your garden by Bougainvilleas. *Indian Bougainvillea Annual,* 25:26-32.

Roy, R.K., Rastogi, R.R. and Singh, S. 2014. Some outstanding Bougainvillea varieties with leaf variegation for ornamental gardens. *Floriculture Today,* 19(2): 24-26.

Roy, R.K., Singh, S., Rastogi, R.R. and Verma, S. 2015. Development of New Varieties of Bougainvillea: The Contribution of CSIR-NBRI. *Indian Bougainvillea Annual,* 26:11-15.

Roy, R.K., Singh, S. and Rastogi, R.R. 2015. Bougainvillea – Identification, Gardening and Landscape Use. CSIR-NBRI, Lucknow, pp.144.

S.S.Sindhu and Roy, R.K. 2008. Year round calendar for Bougainvillea and varietal selection. *Bougainvillea Annual* 21: 12-14.

Salimei, S. 1960. The genus *Bougainvillea* and its species most suitable for growing in Italy. *Italia agric.* 97: 551-5.

S.C., Sharma and Roy, R.K. 2002. Conservation and improvement of Bougainvillea at NBRI. *Floriculture Today* VI (8): 8-9.

Sharma S.C., Roy, R.K. and Srivastava, Richa. 2005. Genetic improvement of Bougainvillea by mutation breeding. *Bougainvillea Annual* 18:8-10.

Sharma S.C., Roy, R.K. and Srivastava, Richa .2005. Role of Bougainvillea in Mitigation of Environment Pollution. *Environ. Science & Eng.* 47(2):131-134.

Sharma, M. D. 1962. The Bougainvilleas introduced in 1961. *Lal Baugh J. Mysore hort. Soc.* 7(1): 18-23.

Sharma, M. D. 1963. Some bewitching Bougainvilleas from Lal Bagh. *Indian Hort.* 7(3): 19-20.

Sharma, M. D. 1968. Know this *Bougainvillea* 'Million Dollar'. *Indian Hort.* 13(1):26.

Sharma, S. C. 1969. A new sport of *Bougainvillea. Indian Hort.* 13(3): 20-31

Sharma, S.C. 1986a. *Bougainvillea* in variegated forms. Newsletter *Bougainvillea* Society of India, 6(1): 11-14.

Sharma, S.C. 1986b. Horto-taxonomical studies on Bougainvilleas. Ph.D. Thesis Kanpur University, Kanpur.

Sharma, S.C. 1996. Bougainvilleas in India. EBIS, NBRI, pp. 74.

Sharma, S.C. and Goel, A.K. 2006. Horto-taxonomy of ornamentals: Case study of *Bougainvillea. J. econ. Boi.*, 30(1) : 71-78.

Sharma, S.C. and Roy, R.K. 2000. Bougainvillea India's Germplasm Collection. *The National Plant Collection Directory 2000.* National Council for the Conservation of Plants & Gardens (United Kingdom): 61-62 & 73.

Sharma, S.C. and Roy, R.K. 2001. Conservation and improvement of Bougainvilleas. *Bougainvillea Annual*, 16: 9-12.

Sharma, S.C. and Roy, R.K. Conservation and improvement of *Bougainvillea* at NBRI. *Floriculture Today*, 2002, 6(8): 8-9.

Sharma, S.C. and Srivastava, S. 1988. Million dollar Bougainvillea, *Newsletter, Bougainvillea* Society of India, 7(1): 1-3.

Sharma, S.C. and Srivastava, S. 1989. Bicoloured Bougainvillea. *Bougainvillea* Society of India *Newsletter*, 8(1): 7-11.

Sharma, S.C., Goel. A.K. and Roy, R.K. 2008. Conservation and documentation of Bougainvilleas in NBRI Botanic Garden in *Proceeding of National Conference on Bougainvillea*, 2006, EBIS, NBRI, p: 6-14.

Sharma, S.C., Roy, R.K. and Basario, K.K. 1991. Bougainvillea as cascade. *Bougainvillea Newsletter* 10 (1): 7 - 10.

Sharma, S.C., Srivastava, R., Datta, S.K. and Roy, R.K. 2002. Gamma rays induced bract colour mutation in single bracted Bougainvillea 'Palekar'. *Journal of Nuclear Agriculture Biology*, 31 (3-4): 206-208.

Sindhu, S.S., Roy, R. K. and Verma, A.K. 2012. Bougainvillea – Classification and its varietal wealth. *Indian Bougainvillea Annual*, 24: 48-53.

Singh, B. and Dadlani, N.K. 1986. *Bougainvillea.* In: Ornamental Horticulture: in India (Eds. Chaddha & Chaudhury). IARI, New Delhi, 10-12.

Singh, B., Panwar, R.S., Voleti, S.R., Sharma. V.K. and Thakur, S. 1999. The New International *Bougainvillea* Check- List. IARI, New Delhi, 1-7

Singh, S.P. and Motial, U.S. 1979. Propagation of *Bougainvillea* cv. Thimma under intermittent mist - I. *PlantScience.*, 11: 53-59.

Sivapalan, A. and Hamid F.H. 1997. Bacterial leaf spot of *Bougainvillea* caused by *Pseudomonas andropogonis* in Brunei Darussalam. Bulletin OEPP/EPPO Bulletin 27, 273-275.

Sobers, E. K. and Martinez, A.P. 1966. A leaf-spot of *Bougainvillea* caused by *Cercospora Bougainvillea. Phytopathology* 56: 128-30.

Sobers, E.K. and Seymour, C.P. 1968. Distribution, Pathogencity and Taxonomy of *Cercosporidium Bougainvillea,* Florida State Horticultural Society 386:397-401.

Srivastava, R., Datta, S.K., Sharma, S.C. and Roy, R.K. 2002. Gamma rays induced genetic variability in Bougainvillea. *Journal of Nuclear Agriculture Biology* 31 (1): 28-36.

Srivastava, R., Shukla, S., Soni, S. and Kumar A. 2009. RAPD based genetic relationship in different Bougainvillea cultivars. *Crop Breeding and Applied Biotechnology,* 9:154-63.

Standley, P.C. 1931. 'Mrs. Butt'. In: Nyctaginaceae and Chenopodiaceae Northwestern South America. *Fld. Mns. Nat. Hist. (Bot. Ser.),* 11: 96-101.

Swarup, V. and Singh, B. 1964. Pollen morphology and hairs in classification of *Bougainvillea. Indian Journal of Horticulture,* 21: 155--64.

Tse, A., Ramina, A., Hackett, W.P. and Sachs, R.M. 1974. Enhanced inflorescence development in Bougainvillea 'San Diego Red' by removal of young leaves and cytokinin treatments. *Plant Physio.* 54:407-414.

Verma, S.C., Haider, M.M. and Murthy, A.S. 1992. Note on effect of chemicals on rooting in *Bougainvillea. Indian Journal of Horticulture,* 49: 284-86.

William, T.G. 1976. Bougainvilleas for cultivation (Nyctaginaceae). *Baileya,* 20(1): 34-41.

Yan, Wu Xiao. 2012. Studies on genetic diversity and phylogenetic relationship of ornamental germplasm resources in bougainvillea. Master's Thesis, Huaqiao Univ., China

Zadoo, S.N. and Khoshoo, T.N. 1968. Cytogenetical studies in *Bougainvillea*: A case of interchange heterozygosity. *Genetica,* 39: 353-60.

Zadoo, S.N. and Khoshoo, T.N. 1975. Nature of self-incompatibility in cultivated Bougainvilleas. *Incompatibility Newsletter,* 5: 73-75.

Zadoo, S.N., Roy, R.P. and Khoshoo, T.N. 1975a. Cytogenetics of cultivated Bougainvilleas- I: Morphological variation. *Proceeding of Indian National Science Academy,* 41B: 121-32.

Zadoo, S.N., Roy, R.P. and Khoshoo, T.N.1975c. Cytogenetics of cultivated *Bougainvillea* - III: Bud sports. *Z. Pjlanzenzuchtg,* 74: 223-39.

Zadoo, S.N., Roy, R.P. and Khoshoo, T.N. 1976. Cytogenetics of cultivated Bougainvilleas - VII: Origin and evolution of ornamental taxa. *Indian Journal of Horticulture,* 33: 278-88.

Zadoo, S.N., Roy, R.P. and Khoshoo, T.N.1975b. Cytogenetics of cultivated Bougainvilleas- II: Pollination mechanism and breeding system. *Proceeding of Indian National Science Academy,* 41B: 498-502.

Zadoo, S.N., Roy, R.P. and Khoshoo, T.N.1975d. Cytogenetics of cultivated *Bougainvillea*- IV: Meiotic system. La Cellule, 71: 311-22.

Zadoo, S.N., Roy, R.P. and Khoshoo, T.N.1975e. Cytogenetics of cultivated *Bougainvillea*- V: Tetraploidy and restoration of fertility in sterile varieties. *Euphytica,* 24: 517-24.

Zadoo, S.N., Roy, R.P. and Khoshoo, T.N.1975f. Cytogenetics of cultivated *Bougainvillea*- VI: Hybridization. *Z. Pflanzenzucbiq.* 75: 114-37.

Sources of Bougainvillea Plants in India

1. Adwait Nursery, Paranwadi, Pune , Maharashtra.

2. Arambagh Nursery, Pallishree, Arambagh, Hooghly, West Bengal.

3. Botanic Garden Division, CSIR-national Botanical Research Institute, Rana Pratap Marg, Lucknow-226001, Uttar Pradesh.

4. Classic Farms & Nurseries, Regd. Office: G-3, Ashraya Apartments, 12-7-273, Mettuguda, Secunderabad – 500 017, Andhra Pradesh.

5. DBN Agro-Horticulture (P) Ltd., Regd. Office: Vill – Sripur, P.O.- Sripur Bazar, Dist – Hooghly – 712 514, West Bengal.

6. Division of Floriculture, Indian institute of Horticulture Research, Hessarghatta, Bangalore -560089, Karnataka.

7. Ever Green Nursery, Karbala, Jor Bagh Road, New Delhi – 110 003.

8. Floriculture & Landscaping Division, Indian Agricultural Research Institute, Pusa, New Delhi- 110012

9. Friends Rosery, B-110, Mahanagar, Lucknow, Uttar Pradesh.

10. Gardenia Plants Nursery, Plot No. 90, Floriculture Park, MIDC, Talegaon, Pune, Maharashtra.

11. Govt. Sunder Nursery, Nizamuddin, New Delhi - 110 013.

12. Horticulture Development Centre, Judge Court Road, Mominpur Kolkata – 7001 027.

13. Indo-American Hybrid Seeds (I) Pvt. Ltd., 17th Cross, K.R. Road, Banashankari II Stage, Bangalore – 560 070, Karanataka.

14. Jagtap Nursery's Garden Centre, Magarpatta, Pune, Maharashtra 411013

15. Kamal Nursery, Andul-Mouri, Howrah, West Bengal.

16. Khushboo, 206, Nana Path, Pune – 411 002, Maharashtra.

17. Krishi Kranti Kendra, Shani Mandir Road, Sitabuldi, Nagpur, Maharashtra.

18. Krishnendra Nursery, 9-159, Lalbagh Siddapura, Jayanagar 1st Block, Bangalore – 560 011, Karnataka.

19. KSG & Sons, 177, 5thMain road, Chamrajpet, Bangalore – 560 018, Karnataka.

20. Masjid Nursery, Pandara Road, New Delhi – 110 092.

21. Namdeo Umaji Agritech (I) Pvt. Ltd., 1205/4, Alankar,1st Floor, Opposite Sambhajii Park, Shivaji Nagar, Pune – 4, Maharashtra.

22. New Janta Nursery, Ludhiana Road, Malerkotla- 148023, District - Sangrur, Punjab.

23. Poinsettia Nursery, Pune-Solahpur Road, Pune, Maharashtra

24. Patel Farm & Nursery, Udvada (R.S.), Valsad, Gujarat.

25. Pushpam Nursery, E-18, Liberty Colony, Sarvoday Nagar, Lucknow, Uttar Pradesh.

26. Rajdhani Nursery, Karbala, Jor Bagh Road, New Delhi-110003.

27. Sahayog Hortica Pvt. Ltd., Bakhrahat Road, Samukpota, P.O.; Kanganberia, District: 24 parganas (S), West Bengal 743503.

28. Sandeep Nursery, B-69, Mandir Marg, Mahanagar, Lucknow, Uttar Pradesh.

29. Sri Ram Nursery, 1-10-35, Behind Rahul Automobiles, Begumpet, Hyderabad – 500 016

30. The Agri-Horticultural Society of India, 1, Alipore Road, Kolkata – 200 027, West Bengal.

31. The Deb Narayan Nursery, Sukhdevpur, Bishnupur – 743 503, South 24 Parganas, West Bengal.

32. The Mondal Nursery, Bakhrahat, 24 PGS (S), West Bengal.

33. The Nurserymen Co-operative Society Ltd., Lalbagh, Bangalore – 560 004, Karnataka.

34. The Rainbow Garden, Mahamayapur, 24 PGS (S), West Bengal.

35. The Indian Nursery, Andul, Howrah, West Bengal.

36. The Agri-Horticultural Society, 31, Cathedral Rd, Near Semmoezhi Ponga, Gopalapuram, Chennai-600086, Tamil Nadu.

Subject/Terminology/Index